THE
HORSELESS
CARRIAGE

THE BIRTH OF THE MOTOR AGE

THE
HORSELESS
CARRIAGE

THE BIRTH OF THE MOTOR AGE

STUART HYLTON

For Sheila

'We declare that the world's wonder has been enriched by a fresh beauty, the beauty of speed. A racing car with its trunk adorned by great exhaust pipes like snakes with explosive breath…is more beautiful than the Victory of Samothrace….'

From *The Futurist Manifesto*, Filippo Tomassa Marinetti, 1909

First published 2009
Reprinted 2020

The History Press
97 St George's Place, Cheltenham,
Gloucestershire, GL50 3QB
www.thehistorypress.co.uk

© Stuart Hylton, 2009

The right of Stuart Hylton to be identified as the Author
of this work has been asserted in accordance with the
Copyrights, Designs and Patents Act 1988.

British Library Cataloguing in Publication Data.
A catalogue record for this book is available from the British Library.

ISBN 978 0 7524 5078 0

Typesetting and origination by The History Press
Printed in Great Britain by TJ International Ltd, Padstow, Cornwall.

CONTENTS

INTRODUCTION

In my book *The Grand Experiment* (Ian Allan 2007) I told the story of the first twenty-five years of the railway age. It documented the dramatic changes that railways made to our perception of time and space across the entire world, and the impact of railways on the economy and society. In the course of writing it I was struck by the fact that each of the major transport developments of the past two centuries had their own revolutionary impacts in their first few decades. In the case of aviation, this takes us from the Wright Brothers to Guernica in the Spanish Civil War and the prototype Spitfire, and the profound impact that aviation was to have on the conduct of warfare – the full effects of civil aviation were yet to be felt.

With the motor car the impact was felt in terms of personal mobility. Important though they were, railways opened up areas within a relatively narrow proximity to the railway stations. Even in a small and densely developed country like Britain the railways could not be everywhere. Beyond the stations the pace of life was still dictated by the speed at which men could walk or horses could carry them. For reasons which this book explores, the clear potential for steam traction to transfer across from rail to road was never realised.

It was the boom in bicycling in the 1870s that first gave the public in Britain and abroad a taste of a new era in personal mobility. But it was the motor car that was to reveal its full potential. Over a period of about a quarter of a century, from the first limited production of cars in 1891 to the outbreak of the First World War, there was a revolution in the performance, reliability and affordability of the motor car, which changed the course of the twentieth century just as railways changed that of the nineteenth. Huge areas of the country were exposed to tourism, became potential commuter areas, and focal points for a host of other economic activities. In our towns and cities, one health hazard (the massive amounts of horse manure in our streets and the diseases which came with it) led to the motor car being promoted as a major contributor to clean air. One thing that did not change: traffic jams remained an urban fact of life, whether the traffic was horse-powered or driven by petrol.

In this book I have tried to provide an overview of the development of the motor car up to 1914, and its impact on society. There is nothing magic in the choice of 1914 as an end date, beyond the following facts: that most of the world became preoccupied by other weightier matters than the development of the motor car from that date on; that much of the manufacturing capacity and technical expertise of the emerging motor industry was diverted

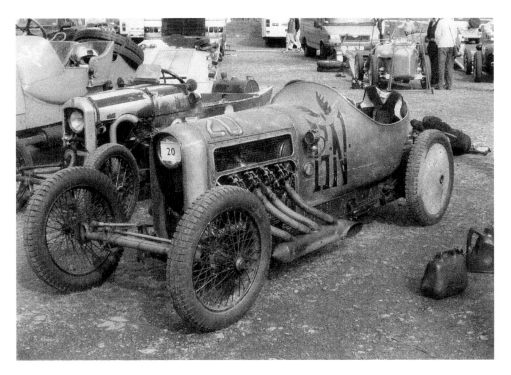

Between 1910 and 1925 the London company GN (named from the initials of its founders H.R. Godfrey and Archie Frazer-Nash) built some of the most popular cyclecars. This much more substantial 5.1-litre JAP-engined beast dates from 1908. Frazer-Nash later built cars under his own name.

to the war effort; that international cooperation – and sporting competition – both of which contributed to technical development, largely came to an end (though war can be its own driver of technological change, and the conflict gave the internal combustion engine a first real chance to demonstrate its military applications).

In fact, as others have argued, the cars of 1914 had much more in common with those of the 1930s than they did with their 1904 counterparts, in terms of their performance and equipment. In some respects 1904 (the date that separates veteran from vintage cars in Britain) might have been a better end date to choose. By then, many of the early blind alleys of motor car technology – such as tiller steering, hot-tube ignition, belt drive, horizontal engines and the possibility of steam as an alternative to internal combustion – had been largely abandoned and the broad layout of the modern car established. Largely abandoned too was the practice of inter-city racing on public roads, following the horrendous accident rate that had its unhappy climax in the 1903 Paris–Madrid race. The age of the closed, purpose-built racing circuit was about to dawn, led by Brooklands in Britain and Indianapolis in the United States. But the period I have chosen does cover the most interesting phase of the development of motoring, as society struggled to get used to the idea of individual members of public travelling about among us at what were previously seen as supernatural speeds, while the motor industry wrestled with the technology that enabled them to do so safely and reliably.

Different countries went down different routes, and at different speeds, on entering the age of the motor car. In the United States the development of car technology proceeded rather more slowly than in Europe, though American manufacturers led the way in developing modern manufacturing techniques.

In writing this book I am keenly aware that there are others far better qualified than myself to cover different aspects of the topic. In particular, I make no claims whatsoever to be a mechanical engineer, and these parts of the book have been as much a voyage of discovery for me as they will be for some of my readers. At the same time, I hope that my layman's attempt to explain the mechanical development of the motor car up to 1914 will be of interest to, and not beyond the grasp of, the most technologically challenged of my readers. If I can understand it, anyone can!

I would like to acknowledge a debt of gratitude to all the sources listed in my bibliography. In particular (in alphabetical order):

- Cauntier and his account of the evolution of the light car;
- Georgano's book explaining early coachbuilding technology and the blind alley of following traditional horse-drawn methods and designs;
- Knight's 1902 book on motoring, giving a contemporary perspective on the activity;
- The books by Nixon, and Karslake and Pomeroy, for their fascinating insights into the experience of being an early motorist;
- Newcombe and Spurr, for their authoritative but accessible account of how early motor car technology evolved;
- Nicholson's comprehensive three-volume account of the evolution of the British motor car; and
- Villard's fascinating (if sometimes horrifying) history of the early road races.

I commend these and all the other sources listed in my bibliography to any reader who would like to find out more. Any errors or omissions in the book, despite this wealth of authoritative guidance, are entirely my own responsibility. In some cases consideration of space forced me to curtail the detail or scope of my coverage. In particular I had to lose an entire chapter devoted to the various attempts (through voiturettes and cyclecars to public transport) to make powered road transport more widely available.

The historic photographs in the book come from the collection at the National Motor Museum, Beaulieu, and my thanks go to Jonathan Day, the Picture Library Manager, for his help in tracking them down. I have tried to identify and seek permission from any copyright holders, but if any have slipped the net, please let me know via the publisher and I will try to ensure that this is rectified in future editions.

Last but not least, a word of thanks to my wife Sheila, who has patiently endured many years of my obsessive scribbling. P.J. O'Rourke once dedicated a book to his wife with the words 'Thank you for not killing me, dear', to which I would only add 'so far'.

Stuart Hylton
October 2008

CHAPTER 1

WHAT IS IT?

What is it?

It is an autocar.
Some people call it a motor car.
It is worked by a petroleum motor.
The motor is of four horse-power.
It will run sixty miles with one charge of oil.
No! It can't explode – there is no boiler.
It can travel at fourteen miles per hour.
Ten to eleven is its average pace.
It can be started in two minutes.
There are eight ways of stopping it, so it can't run away.
It is steered with one hand.
Speed is mainly controlled by the foot.
It can be stopped in ten feet when travelling at full speed.
It carries four gallons of oil and sixteen gallons of water.
The water is to keep the engine cool.
It costs less than ¾*d* a mile to run.
The car can carry five people.
It was built by the Daimler Motor Company of Coventry, and cost £370.
We have come from John o'Groats House.
We are going to Land's End.
We are not record breaking, but touring for pleasure.

(Taken from a card produced by Henry Sturmey for his October 1897 drive from John O'Groats to Land's End, in answer to the most commonly asked questions about his vehicle. His car was only the second British-built Daimler ever to be made.)

CHAPTER 2

LIFE BEFORE THE MOTOR CAR

THE ROADS

If the motor car was ever to be a practicable proposition, it would need a decent road network. The roads in late nineteenth-century Britain were at the end of a period of decline lasting many centuries and punctuated only by a relatively brief renaissance in their fortunes. Little real investment had gone into most British roads since the Romans had departed in the fifth century.

An Act of 1555 had assigned responsibility for the upkeep of the network to the individual parishes through which they ran. Two unpaid surveyors were to be appointed to oversee repairs and each able-bodied parishioner was required to donate four (later six) days' unpaid labour a year to road-mending. The system was universally hated, widely ignored and almost impossible to enforce. The result was that, by 1700, most of the Roman road network could barely be detected, far less used.

The period of renaissance was the turnpike movement, in which the construction of improved roads was funded by tolls paid by their users. This started with works along parts of the Great North Road in 1663 but did not really get under way until the 1720s. An additional factor stimulating road building at that time was the Jacobite rebellions of 1715, 1719 and 1745, which prompted the construction of miles of roads for military purposes in the highlands.

Turnpiking really came into its own in the second half of the eighteenth century. Between 1751 and 1791 some 454 new turnpike bills were passed, in a period known as 'turnpike mania' where the mileage of turnpiked roads was roughly quadrupled. However, the turnpikes were almost completely unregulated, beyond their initial sanction from Parliament, with the result that the standards of construction and maintenance were hugely variable. Many of the surveyors were still amateurs and part of the labour force consisted of the same unskilled and unwilling people who had been turned out to do unpaid work under the 1555 statute.

But good, bad or indifferent, even at their height in about 1835 turnpike trusts only managed about a fifth of the national road network – about 22,000 miles of road. Up until that time the remainder was still maintained – if they can be dignified by such a term – by the primitive parish system. The formation of County Councils in 1888 proved to be a step forward in delivering a more consistent, professional approach to highway maintenance. But, before that could happen there was yet another period of decline. The spread of railways meant that the turnpikes lost much of their trade and income. Many trusts went bankrupt and such highway maintenance as they provided was lost with them. But, whilst these roads may have lost much of their town-to-town traffic to the railways, the overall national volume of road-based commercial traffic increased as road haulage companies sought business that complemented, rather than competed with, the railways. During the nineteenth century commercial road vehicles increased from 116,000 in 1811 to an estimated 161,000 in 1851 and 702,000 in 1901.

For much of the nineteenth century, however, roads were something of a Cinderella mode of transport, while the railways boomed and attracted all the available private investment. However, from the 1870s onwards pressure for better roads came from the start of the aforementioned craze for cycling. This was to benefit the car industry in two ways. First, it created a host of small manufacturing firms, a remarkable number of whom would be among the pioneer producers of motor cars. Second, it created an impetus for improved highway maintenance.

As for the construction of roads, it was the early nineteenth century before we saw any important developments. Thomas Telford devised a method of building roads with extensive foundations, but this was too expensive to have a wide application. More significant was the man who was appointed surveyor to the Bristol Turnpike Trust in 1815. John McAdam developed a new method of constructing roads which was relatively hard-wearing, easy to maintain and – most importantly – cheap to build. It could be laid directly onto a suitable (level and well-drained) sub-soil, and consisted of a surface of compacted gravel and broken stone, bound together by clay, and cambered to improve the drainage. By 1824 some 15 per cent of the turnpike network was in the process of being macadamised.

Many cycle manufacturers turned their hand to car-building. With some, like this 1898 Stephens dogcart, built near Bristol, the origins of the firm were more obvious than others.

A graphic illustration from 1903 of the problems caused by cars on dusty roads.

Although his method was a great improvement on what had gone before, McAdam's roads were still prone to rutting and were very dusty in dry weather. As long as the only heavy vehicles moving along them were slow, they could perform relatively well, taking four to five tons, distributed over four wheels. But the coming of faster, albeit lighter, motor cars created a new challenge. Their pneumatic tyres apparently used to suck out the clay binding material of the road, creating a huge amount of dust and causing the road surface to disintegrate.

The solution to these problems was not discovered until well into the car age, and then only by accident. It was in 1901 that the county surveyor of Nottinghamshire, a man named Edgar Purnell Hooley, spotted that a particular stretch of one of his roads was free from ruts. It turned out that a barrel of tar had accidentally been spilt on it, and that slag from a local ironworks had been used to cover the spillage. Within a year Hooley had patented the process under the name 'tar macadam', and went on to establish the Tar Macadam (Purnell Hooley's Patent) Syndicate Ltd. The process was later registered under the rather more digestible trademark of 'tarmac'. Georges Haussmann, the planner of nineteenth-century Paris, had previously introduced asphalt road surfaces for rather different reasons. There it was used to deny a rebellious populace a supply of cobblestones for throwing at the authorities or using to construct barricades.

So, travelling by road could be a dirty business at the dawn of the car age. If Britain was starting the motoring age with a backlog of neglect and under-investment in its roads, things were different in parts of continental Europe, France in particular. Again, the original motivation was military. Whereas Britain relied for its defence on the seas that surrounded it, the nations of continental Europe had vulnerable land borders. It was important for them to be able to move troops quickly to wherever there was a threat. As early as 1716 the French founded a body called the Corps des Ponts et Chaussees, who had responsibility for improvements to the highway network. By 1776 half of France's roads were being rebuilt and realigned and a system of *routes nationales* had been classified. Napoleon was, not surprisingly, very keen on the programme. Napoleonic France even had her own McAdam – a man called Pierre Tresauguet, whose road construction techniques were adopted as a national standard.

But if things were bad in Britain they were even worse for motorists in America. As of 1900 there was reckoned to be just 160 miles of surfaced highway in the whole of that vast nation. Beyond that, according to those who tried to drive through it, conditions in rural America were even worse than anything the first long-distance motorists faced in the wilds of Asia. The Model T Ford, which did so much to open up the rest of the nation to the possibilities of motoring, relied much more on its ruggedness than its sophistication for its appeal to rural America.

As a footnote, anticipating the rest of the book, the motorway – a purpose-built high-speed car route – has a much longer heritage than the pre-war German autobahns, and dates back to the period in which we are interested. The prototype was William K. Vanderbilt's Long Island Expressway, dating from 1906-11, described in a later chapter. Plans for a similar route between London and Brighton were being contemplated at about the same time, but the RAC were fearful that the building of express roads for cars would lead to them being banned from the general road network, and the plan was dropped. Lord Montagu put forward proposals for the modern equivalent to a turnpike in 1923 – a toll motorway linking London, Birmingham, Liverpool and Manchester. But it was opposed by both parliamentary and private interests, and again the idea came to nothing.

EARLY TRAFFIC: HORSE POWER AND STEAM POWER

So what vehicles were travelling along the roads in the years before the motor car made its appearance? There are records of *long wagons*, four-wheeled passenger vehicles carrying up to twenty-five people, as early as the sixteenth century, but in those days walking (or, if they could afford it, travelling on horseback) was far preferable to any vehicle for most people.

Although the first stagecoach ran as early as 1640, the ownership of a carriage in Stuart times still remained more of a status symbol than a practical means of transport once you got much more than thirty-five miles out of London. The state of Britain's roads continued to mean that there was no strong commercial incentive to develop the design of horse-drawn vehicles. Only the turnpike movement made the relatively easy movement of carriages and carts a realistic possibility. But travelling by coach was for the wealthy elite. For most who could afford any form of transport other than walking, the choice would have been the carrier's wagon, or *stager*. Although this cost only ½d a mile, it lumbered along at a mere 3mph (or fifteen miles a day), was frequently diverted to make deliveries, and you had to get off and walk when going uphill. According to Newcomb and Spurr it was so tedious that 'none but women and people of inferior condition travel in this sort.'

The introduction of faster, lower and lighter mail coaches from 1784 led to improvements in the design and construction of horse-drawn vehicles generally, with developments like 'shrunk-on' iron tyres and steel springs appearing during the late eighteenth century. Networks of coaching inns grew up around the country offering a slick change of teams of horses (a changeover time of forty-seven seconds was claimed on the Devon mail coach *Quicksilver*). This, and a willingness to drive the horses to the point which reduced their working lives to about two years, meant that average journey speeds of 12 to 14mph could be achieved.

This was very much at the physical limit of what horse-drawn travel could achieve. The high-point of the coaching era came to an end with the opening of the first railways. Those coaching services in direct competition with railways rapidly went under, while the rest of the industry adapted, often by providing complementary feeder services to the railway stations.

The craftsmanship of English coachbuilders was recognised to be among the finest in the world, and the industry's products were exported to virtually every country that had roads fit to carry them. But this, if anything, disadvantaged the development of the car industry in Britain.

Horse-drawn vehicle design was, as we shall see, a long way from being able to meet the new demands that the internal combustion engine would place on the vehicles it powered.

Perhaps most surprising was how far horse-drawn public transport survived into the motor car age. The last stagecoaches were built in 1910, the final attempt at a daily coach service between London and Brighton only closed in 1925, and the last recorded public passenger service using a coach and four horses for anything other than tourism came as late as 1933.

THE HORSELESS CARRIAGE

But if horse-drawn travel was a long time dying out then the gestation period for the horseless carriage was even longer. If we take the literal meaning of the term *automobile* to mean any vehicle capable of providing its own power, the idea has its origins some 2,000 years before the first petrol-driven cars took to the road. Hero of Alexandria foresaw the possibility of some kind of steam-powered vehicle supplanting horses as long ago as 150 BC, and Leonardo da Vinci sketched some car-like designs in his *Codice Atlantico*. These included a design for a differential gear well before there was any obvious need for such a device. Jellinek-Mercedes quotes two early, if not particularly conclusive, accounts of horseless carriages:

> The *Memmiger Chronicle* mentions, according to Feldhaus in his book *Ruhemesblatter der Technik*, the trial on 2 January 1447 of a carriage 'without horse, oxen or people, yet the man who built it sat in it'.

And in the *Chronicle of Pirna* by Petrus Albinus:

> Strange carriage: in the year 1504 a citizen constructed a carriage with wheels and all kinds of springs, which could drive it without a horse, though one person sat on it and worked it. He attempted to drive it two miles towards Dresden but he did not get very far for he stuck in the mud which was at the time very thick; on the flat and dry he could have made it. There were many people present as everybody wanted to see this thing. (Jellinek-Mercedes, pages 30-31)

Two optimistic gentlemen by the names of Ramsey and Wildgoose took out a remarkably vague patent in 1618 for 'newe, apte or compendious formes or kinds of engines or instruments to plough grounds without horse or oxen.' No indication survives of what – if anything – they had in mind to make the device, whatever it was, move!

Father Ferdinand Verbiest was a Belgian missionary attached to the Court of the Emperor of China. Part of his efforts to convert the court to Christianity involved dreaming up ingenious devices and, in 1672, he came up with a toy which moved by blowing steam onto the vanes of a turbine. It had a very limited range, was not large enough to carry people (being a scale model about 60cm long), and there is even some doubt as to whether it was ever actually built. These minor considerations apart, it can lay claim to being the first working steam-propelled vehicle. Shortly after this, in 1690, a French physicist called Denis Papin produced the first condensing steam engine.

This was further improved upon in Britain by Newcomen and Crawley and, more importantly, by James Watt who, in 1769, patented a separate condenser system and, in 1784, secured another patent that sought to cover every aspect of steam power. This included a high-pressure steam-propelled carriage (something that he himself had no intention of building). Watt's stationary engines went into successful production, powering the emerging industrial giant that was eighteenth-century England. Their large size, slow speed and inefficiency meant that they were only appropriate for stationary applications, and Watt's robust defence of his patent (he spent £30,000 doing so) inhibited others from developing new applications of steam to power vehicles. This cost Britain the lead in this form of steam power.

The competition overseas was less constrained. Oliver Evans of Pennsylvania obtained licences to operate steam-powered road wagons in 1787 and, three years later, Nathan Read ran a steam carriage on the streets of Philadelphia. In Europe the French military engineer Nicolas Joseph Cugnot built a curious-looking vehicle in 1763 (or '65). It carried four people at speeds of up to 5.9mph. This was followed by a military prototype designed to tow cannon of up to four tons in weight. However, it was painfully slow (2.2mph), hopelessly lacking in range (with a running time of about fifteen minutes, or half a mile, before it had to stop for another fifteen minutes to rebuild pressure) and it was (to the surprise of no modern reader who looks at it) almost impossible to steer, combining as it does drive and steering through a single wheel. Its test report read: 'The excessive violence of its movements prevented its being steered and, as early as its first trial, a piece of wall which happened to get in its way was knocked down.'

THE FIRST PRACTICAL STEAM CARRIAGES

A carriage has been constructed containing a small steam engine, the force of which was found sufficient, upon trial, to propel the carriage, containing several persons, amounting to at least a ton and a half in weight, against a hill of considerable steepness, at the rate of four miles in an hour. Upon a level road, it ran at the rate of 8 or 9mph. (*Camborne Gazette* and *Falmouth Packet*, 20 February 1802)

Watt's patent expired in 1800 and on Christmas Eve 1801 Cornishman Richard Trevithick first ran his steam-powered road-going vehicle, described above. It was quite a sophisticated machine for its time, anticipating developments like the blast-pipe exhaust that would only become standard practice on railways decades later. Carrying seven passengers, it ascended the 1:20 incline on Camborne Beacon in West Cornwall 'like a bird'. It proved to be capable of 9.3mph on the flat and 3.7mph uphill. The vehicle worked perfectly well until it broke down, fortuitously, outside licensed premises. Trevithick and his companions managed to manoeuvre the machine into a lean-to by the inn, whereupon:

…the parties adjourned to the Hotel, and comforted their Hearts with a Roast Goose and proper drinks, when, forgetful of the Engine, its Water boiled away, the Iron became red hot, and nothing that was combustable remained either of the Engine or the house. (Contemporary account, quoted in Brandon, page 16)

Another model was developed which was driven to and around London, but nobody showed any commercial interest in it. On the contrary, it was said that some ran in terror from the belching monster, referring to it as 'the Puffing Devil'. It was eventually scrapped.

The Napoleonic Wars inhibited research into steam propulsion and, in the years that followed, it was the development of railway locomotives that tended to get more attention than their road-going cousins. However, steam-powered road vehicles at first evolved at a similar rate to those on rail. By the early 1820s a number of experimental steam coaches were operating in Britain. Some, such as those designed by John Tindall (1814) and David Gordon (1824), even had ingenious systems of 'legs' and 'feet' that enabled the carriage to 'walk', but these proved to be too mechanically complicated to be reliable and did too much damage to the road surface.

William Henry James tested a steam-powered carriage in 1829. It could carry twenty or more people and reached speeds of 15mph. But as with so many of its contemporaries, the engineering of the day struggled to keep up with the demands being made on it, and the machine could rarely go more than three or four miles without something breaking. Lack of finance prevented James from refining his invention.

Goldsworthy Gurney was one of the leading steam carriage pioneers of the day, and in 1828 one of his vehicles managed a fully laden ascent of Highgate Hill to Barnet. However, on the

This steam coach claimed to offer journeys at 14mph for 3*d* a mile.

way back down the vehicle's brakes failed, sending it crashing at high speed into a ditch. The intended purpose of the carriage was to provide a regular service between London and Bath. It had to stop every four miles or so for water, finally broke down and suffered the indignity of being towed into Bath by a team of horses, but it completed the return journey successfully in ten hours. But Gurney couldn't find any backers for his carriage. The following year, he produced a rather more practical steam tractor unit for a coach, which hauled a barouche containing 'the Duke of Wellington and other Persons of Distinction' through Hounslow.

He built other machines over the following years, one of which provided a regular service between Cheltenham and Gloucester (1831), and another which managed a return trip between London and Brighton in a total journey time of ten and a half hours. Unsuccessful attempts were made to obtain Government funding for Gurney's work. Equally unsuccessful were the attempts, discussed later, to secure the repeal of the Turnpike Act, which allowed turnpike trusts to set their own – often prohibitive – tolls on steam-powered vehicles. Twice the Steam-Carriage Bill foundered in the face of opposition in the House of Lords, causing a discouraged Gurney to abandon steam transport for other fields of invention.

Messrs Summers and Ogle built a steam carriage that was said to have been both reliable and fast. It averaged 10-12mph on a journey between Oxford and Birmingham, and was claimed to be capable of up to 32mph. But perhaps the most ambitious effort of all came from a Dr Church of Birmingham, who built a massive fifty-seat steam carriage that was intended to provide a regular service between London and Birmingham. Unfortunately it broke down during its second trial run and never entered service.

By the 1830s there were about 100 steam carriages operating in Britain, with twenty-six in the London area alone, and more than ten companies were operating a steam coach network that reached as far as Holyhead. The leading light among them was Walter Hancock of Stratford, then a village just outside London. In 1830 he produced *The Infant*, a covered charabanc seating ten passengers. From 1831 onwards this provided a regular service between Stratford and London until one of his stokers, Richard Outridge (the French term for stoker, *chauffeur*, was already in use on these vehicles) discovered that he could get more performance out of the machine by wiring down the safety valve. The boiler duly blew up with fatal consequences for Mr Outridge. Hancock undertook longer-distance journeys, such as one from London to Marlborough in 1835 (the same year that the Great Western Railway was given parliamentary approval). Because of a mechanical problem the tender on which they were to have carried their supplies of coke and water had to be left behind, with the result that the trip took on something of an improvised air:

> No one who has not travelled by steam carriages can imagine the inconvenience and delay which result from the want of regular and ample supplies of water, the carriage having to stop for fourteen or eighteen minutes every ten or twelve miles to fill the tanks by hand buckets from pumps, with sometimes the additional inconvenience of having to take the supply from some neighbouring stream or pond.
>
> While the carriage is stationary the fire slackens, in consequence of the blower being stopped, and it requires about two miles running to get it again into full play. By observations which I made on the road while timing the carriage, I found that the rate of the first three miles after taking in water averaged seven and a half minutes to the mile, whilst the latter part of the distance till the carriage again stopped for water averaged one mile in five minutes. Frequently the men were obliged to use any kind of water they could get, some being filled with duckweed, straw and filth of every description, which, of course, very much retarded the generation of steam. (*Mechanic's Magazine*, August 1835)

Hancock went on to produce a range of other steam carriages. Some, such as the *Enterprise*, even looked somewhat like a modern coach and boasted advanced features such as chain drive, a chain grate and a multi-tube boiler. Others incorporated a differential gear. *Enterprise* began running on the route between Moorgate and Paddington that had been taken by the first horse buses just four years earlier. It was a reliable vehicle, and the horse bus operators saw it as

Goldsworthy Gurney built a series of steam carriages from 1828 onwards. His attempt to provide a regular service between London and Bath lasted just one trip.

Walter Hancock operated a series of steam coaches in the 1830s. *Enterprise* provided a regular service between Moorgate and Paddington.

a threat to their livelihoods. They tried to impede it on its journeys which, given the relative weights of Hancock's machine and theirs, seems like a very bad career move for the horse buses. He was not the only one to face such physical opposition. Over in France Charles Dietz tried to establish a scheduled steam carriage service between Bordeaux and Lisbourn, but he met with such violent hostility from local carriers that, on one occasion, he had to fight them off with his poker.

Hancock was eventually forced to retire from the road carriage business. He operated a fleet of ten steam buses in London, but by about 1846 they had ceased to be profitable and he invested instead in the railways. His fate is typical of many of his fellow steam carriage pioneers. The steam road car had by 1840 reached a stage of development (in terms of things like reliability and quiet running) that the petrol-engined car would not match until after the turn of the century.

So why did steam carriages fail to catch on when steam railways were so successful? One reason was the railways themselves, which were rapidly being established along the most profitable routes, and took away much of the trade from the turnpikes, undermining the viability of carriages. Although the initial cost of building a railway infrastructure was greater than that of a carriage running on an existing road, its running costs were less, since a railway locomotive is mechanically more efficient than a comparable road-going vehicle. Against this was the fact that steam carriages, as a relatively new and uncommon form of transport, had not yet been spotted as a potential source of taxation by the governments of the day, and so escaped at least that financial burden.

But any advantage the road-going vehicle had from there being an existing carriageway to use was in many cases negated by the punitive tolls charged by turnpike trusts (for example, the Liverpool to Prescott road charged £2 8s for a steam vehicle, compared with just 4s for a fully loaded coach and four horses). The logic of these charges was not entirely clear. The main reason given was based on the argument that steam carriages, with their greater weight and power, did more damage to the turnpike roads than their horse-drawn equivalents. But this flew in the face of evidence (given to various parliamentary committees by witnesses as eminent as Thomas Telford) that the opposite was the case: the wider wheels of the steam carriages did less harm than narrow stagecoach wheels and horses' hooves. Notwithstanding this evidence, the Locomotive Act of 1861 gave the Board of Trade powers to ban any vehicle that was judged to cause excessive wear and tear on the roads.

But underlying all this was a concern by the trusts that steam carriages, being more efficient than the horse-drawn variety, would further reduce their income that was already, in many cases, suffering serious reductions from railway competition. More fundamentally, there was a huge vested interest in horse-drawn travel from the turnpikes, the coaching businesses, fodder providers, horse breeders, harness-makers and so on.

If the tolls were not enough, some trusts stooped to trying to scupper steam carriages by placing physical obstacles in their way on the turnpike. One of the most notorious of these was the decision of the Glasgow Turnpike in 1834 to heap loose stones on the Glasgow to Paisley road as an obstacle to steam traffic. In trying to avoid them, one of the vehicles turned over, causing a boiler explosion that killed five people and led the courts to ban the service. But, in general, these obstructive tactics appear to have affected their customers just as much, if not more, than the steamer's.

The condition of the roads may also have contributed to the fact that there appears to have been a rather higher rate of mechanical failure in road-going steam vehicles than among their railway counterparts. They certainly would have had to deal with many more bumps and potholes, which would encourage breakages.

Safety was the other issue that the steam coach operators would have to contend with in the battle for hearts and minds. They were not helped by boiler explosions like the one described above. In fact, something like one in five of the steam carriage operations in business between 1831 and 1834 suffered explosions, and the legislation before Parliament at this time sought controls over the size and design of boilers to minimise the danger from them – hardly the right climate for seeking a liberalisation of the law. Other factors that told against them were matters of personality; petty rivalries preventing innovators from building on each others' progress, and the fact that many of them were inventors first and businessmen a very poor second. Interest in steam transport on the roads waned. It was just too difficult, and the last steam public service vehicles were taken off Britain's roads in about 1857.

At the same time small steam engines proved their worth to farmers, powering machinery, and by the 1840s these had started to become self-propelled. They did not pose the same threat to the railways or to horse-related industries as passenger vehicles, and because the army also found them of interest they enjoyed a rather less hostile reception. The legislation of the 1860s (discussed later) that effectively killed off steam-powered passenger transport on the roads and later stunted the growth of the British car industry was much less damaging to this type of steam vehicle.

After the developments of the 1830s almost two decades were to pass before any further significant activity was seen. There had been relatively few attempts to build a private carriage during this first period of interest in steam power. A couple were known to have been designed around 1827/28, but neither was thought to have been built. The first practical steam carriage – and possibly the first working 'motor car' – was built in 1835 by Walter Hancock. It was a four-seat steam gig, and looked like a rather bulky phaeton. This was driven in Hyde Park at speeds of up to 12mph, and was apparently a perfectly practicable machine. But the next phase of development saw more of a focus on private vehicles. It was in 1858 that Thomas Rickett of Buckingham began manufacturing – and, for once, actually managed to sell – steam carriages. One was bought by the Marquis of Stafford for use as his private car (though his outings in it were sufficiently rare for them to attract the attention of the local press), and another went to the Earl of Caithness. Unfortunately, this renewed interest coincided with the introduction of the most repressive legislation against mechanical road vehicles, described elsewhere in the book. Another owner, a London distillery owner named Frederick Hodges, had a steam carriage apparently capable of 30mph. He collected six summonses for speeding in six weeks, despite doing most of his driving in Kent at night in an effort to evade the long arm of the law.

Steam traction engines established at least a small and specialised niche in the commercial transport market, but had had little or no impact on passenger road transport, either public transport or of the personal variety, by the time the first motor cars made their appearance on the continent.

CHAPTER 3

INTERNAL COMBUSTION

You cannot get people to sit over an explosion.

(The reaction of one nervous inventor to the idea of internal combustion, taken from *Bak*, page 20)

New! Practical! Motor car, powered by petroleum, benzene, naptha, etc....patented in all industrial countries… entirely supersedes the horse and wagon.

(Part of the text of what is thought to be the world's first car advertisement, for *Benz* in 1888)

THE ORIGINS OF INTERNAL COMBUSTION

The working principle of a gun, forcing a projectile down a barrel by means of an explosion, and that of an internal combustion engine, forcing a piston down a cylinder by the same means, are very similar. It is therefore perhaps less surprising than it might seem that their development has some overlaps.

As early as 1678 a Dutch physicist named Christian Huyghens set out the theory of using gunpowder to activate a piston. Alessandro Volta established the link in rather more practical terms in 1777 when he developed a pistol which replaced gunpowder with a mixture of marsh gas and air which was ignited by an electric spark. Others, less advisedly, tried it with engines that drew their power directly from gunpowder. The engines tended not to be successful and a number of their inventors did not live long enough to work out why. Volta's experiment inspired Isaac de Rivaz of Switzerland to develop the use of gas and air as a source of mechanical energy, which he patented in 1807. He produced an engine using this principle, and fitted it into a carriage which, it was claimed, local people saw working. We therefore might have had at least the makings of the motor car eighty years earlier, but for the fact that de Rivaz was also an officer in Napoleon's army and other pressing engagements took him away from his research. He died in 1828 without taking his experiments further.

But others were looking at the idea of a gas-air mixture engine before 1830, among them Robert Street in England, Frenchman Philippe Lebon and Peter Cooper in America. Two Englishmen named Brown and Wright both obtained patents for such engines, in 1823 and 1830 respectively. Brown went as far as fitting his engine (which was not a true internal combustion engine as we would understand it) into a one-ton vehicle which was then driven 'with comparative ease' up the 1:10 Shooters Hill in Woolwich, Kent, in 1826.

Around this time, too, the possibility of a liquid fuel came closer with the isolation of benzene. In 1841 an Italian, Luigi de Cristoforis, produced a machine that pumped a liquid fuel into an engine. Eugenio Barsanti and Felice Matteucci developed this into a working internal combustion engine in 1856 although, after experiments with various liquid fuels, they reverted to coal gas and air. By 1860 these were being manufactured in small numbers in Zurich.

In France, in the same year, Etienne Lenoir produced his own gas and air engine in France, but it only developed 1.5hp, and was said to be very fuel inefficient and slow-running (100rpm). Some accounts say in 1863 it was driven on an 18km round trip between Paris and Joinville in an hour and a half, though others maintain that the motor could not even move the carriage into which it was fitted. At about the same time Beau de Rochas laid down the principles of the more efficient four-stroke system (some early models were three-stroke) that was to be the basis on which most modern motor vehicles were developed.

The fundamental problem with gas as the fuel for a vehicle was that it was bulky and the technology to cool it, liquidise it and pressurise it into a usable fuel did not exist at the time. Petrol, on the other hand, has weight for weight more latent energy than most high explosives. Benzene, or benzine, as it was originally called, was not immediately obvious to everybody as a fuel. It was more commonly sold by pharmacists as a cleaning agent. Benz used coal gas in his stationary engines, and only stumbled upon the potential of benzene as a fuel by accident, after a Mannheim housewife bought some for cleaning a pair of gloves. She put the liquid too near the fire and blew the house up, killing several people in the process.

An Austrian, Siegfried Marcus, developed a form of carburettor that would turn benzene into an explosive gas in 1860, and by 1864 had installed a form of gas engine into a pushcart. But aspects of his vehicle were not thought through; for example, it used the rear wheels of the cart as the flywheels of the engine, with the result that the rear of the device had to be lifted off the ground to start the engine. So, whilst Lenoir's vehicle would reportedly not move, Marcus's would not stand still, at least not without the engine stalling. Nonetheless, an eye-witness account survives in the 17 March 1901 edition of the *Allgemeine Automobile-Zeitung* of a successful test run of the device in 1865 (albeit only of a few hundred metres). So far as can be established it never ran again and Marcus, something of an intellectual gadfly, moved on to other interests. His car was later to be called in evidence in the famous court case between Henry Ford and George B. Selden, as proof that Selden was not, as he claimed to be, the originator of the motor car.

A decade later Marcus built three other cars, one of which ended up in the Technisches Museum of Vienna. This had many of the features of a modern car. It was four-stroke, had electric ignition, was water-cooled, had an effective carburettor and even boasted a steering wheel (as opposed to a tiller). This car was still capable of being started when Nixon described it in the 1950s. However, Marcus was banned by the authorities from taking his later vehicles on the road, on account of the noise they made, and he lacked the single-mindedness to challenge them on the matter. Although Marcus could lay serious claims to being the founding father of the motor car, a decade before Daimler or Benz, his cars never went into production. In similar vein Edouard Delamere-Deboutteville was said to have produced a workable motor car in 1884. The Delamere-Deboutteville Motor Carriage started out with a two-cylinder four-stroke gas engine, but soon afterwards made the transition to liquid fuels. It apparently made several successful test runs in Fontaine-le-Bourg, France, but again never found its way beyond the prototype stage. This did not prevent the French from claiming their countryman as the inventor of the motor car and celebrating its centenary in 1984.

Julian Hock of Vienna built a practical internal combustion engine – a two-stroke, running on benzene – in 1870. Finally, it may only have been a lack of determination that ruled out a British claim to have invented the modern motor car. One Edward Butler exhibited a drawing of a mechanically propelled vehicle at the Stanley Cycle Exhibition of 1884. It was even protected by a provisional patent in October of that year, but Butler was dissatisfied with the design and did not pursue the matter further for a couple of years 'because the authorities would not countenance its use.' By the time he sought a patent for the improved version (November 1887) and actually got it on the road (the spring of 1888) others had beaten him to a place in automotive history. Meanwhile, in America George Brayton of Boston tested a vehicle powered by internal combustion in 1876, exchanging ideas with Nikolaus Otto when they met in Philadelphia.

There were many others who played a part in the development of the internal combustion motor or its application in road vehicles over this period. In short, the horseless carriage powered by an internal combustion engine was an idea whose time had come. All that was needed was someone with the determination and energy to make it a practical reality.

KARL BENZ (1844–1929)

> A velocipede driven by Ligroin gas, built by the Rheinische Gasmotorenfabrik of Benz & Cie…was tested early this morning on the Ringstrasse, during which it operated satisfactorily.
>
> (Newspaper report in the *Neue Badische Landeszeitung* of July 1886)

Karl Benz was the son of an engine driver, who spent most of his life in Mannheim, Germany. He set up a machine shop in 1872, where he developed metal bending and forming machines and experimented with a two-stroke engine. He got the engine working on New Year's Eve 1879 and became increasingly preoccupied with the idea of building a horseless carriage. He set up Gasmotorenfabrik Mannheim in October 1882 in hope of pursuing his dream, but got no encouragement whatsoever from his business partners, who saw the idea as a dead end. He left them the following January and by the end of the year had set up Benz & Cie, Rheinische Gasmotorenfabrik. His new business partners were rather more sympathetic to Benz's aspirations, though the main business of the company was still supposed to be the manufacture of stationary gas engines.

Benz worked out solutions to many of the problems of car design from scratch. The result was what is now generally regarded as the world's first true modern motor car.

It looks a primitive and strange device to modern eyes – a three-wheeled vehicle that is more like a Victorian perambulator or invalid chair than the ancestor of today's motor cars. But one very modern feature was that the engine and chassis were combined into a single unit. The single-cylinder 984cc engine developed just 0.9hp at 400rpm and propelled the car to 9.3mph. The four-stroke engine was mounted at the rear, on its side, so that the flywheel could run horizontally to prevent gyroscopic action on the steering while cornering (a fear that later proved to be groundless). The water cooling had no pump or radiator, and relied on the principle of the hot water rising and the colder falling for its circulation (even so, the water also tended to boil away, making regular water stops necessary). The drive from the engine was transmitted to the rear wheels by a combination of a belt and chains, providing only a single gear. The clutch consisted of a freely revolving pulley on the driveshaft, onto which the drive-belt could be moved so that the power was no longer transmitted to the wheels. The frame of the car, made of boiler tubes, had elliptical springs at the rear, but no suspension whatsoever for the front wheel. One other modern feature it lacked was Ackermann steering, the details of which are described later. Benz was unaware of this latter invention and, rather than deal with the problems that would result from a primitive centrally pivoted front axle, opted for a single steered front wheel.

Karl Benz, his wife Bertha, their daughter and a colleague, Fritz Held, in a Benz-Viktoria.

The car attracted large crowds whenever it appeared on the road, and Benz was sufficiently concerned that the authorities might use this as an excuse for banning him that he took to making his proving runs at night. He made various improvements to his prototype for the production model, introducing a more powerful engine, wooden wheels in place of the cycle-type ones, a second gear and suspension to the front wheel (or wheels, as they later became).

The building of cars was very much a family affair for the Benzes; Karl's wife Bertha used to pedal her sewing machine treadle, which had been rigged up to charge his accumulators. In fact it was Bertha and their two sons who took the car on its first long-distance trial run. She decided to visit her mother in Pforzheim, some fifty miles from where they lived. Early in the morning she rose and woke her sons, left her husband a note and pushed the car far enough away from the house for him not to be awoken when it started. When it came to hills, Richard steered and mother and elder son pushed. When the leather brake shoes wore out the resourceful Bertha found a shoemaker to supply her with replacements. She bought petrol from chemists' shops and carried out several running repairs. At one point we know she unblocked a carburettor with a hat-pin and replaced worn insulation on a cable with one of her garters. By evening she reached her mother and telegraphed her husband to tell him. His reply was to tell her to mail the engine of the car back to him, since he needed it for another vehicle, and that she and the boys should return by train.

Benz got his patent on 29 January 1886, and he began to develop his car towards a production model. The second model had front suspension, and the Mark III came with a larger engine (1045cc, delivering 1.5hp at 500rpm), two-speed gears and wooden-spoked wheels. It was this latter model that went into production, and Benz sold twenty-five of these between 1888 and 1894, some of them with engines of 1660 and 1990cc. They proved quite hard to sell at first, possibly due to unfamiliarity with the idea of a horseless carriage. Benz's appearance at the Munich Engineering Exhibition of September 1888 bought him only one sale, but an important one, to Emile Roger, who also took out a licence to manufacture Benz cars in France. He was more successful than Benz himself, selling sixty-nine cars up to the end of 1893 – forty-two in France, fifteen in Germany and twelve to other countries.

In the meantime, in 1890, Benz had bought new, more pro-car, partners into the business: Friedrich von Fisher and Julius Ganss. The 1893 Viktoria, as the production model was called, was the first model with four wheels. Its name commemorates Benz's victory over the problem of steering a vehicle with two front wheels. Its engine had increased to 2 litres, but it was still producing just 3bhp. In 1894 a smaller and cheaper model, the Benz Velo, led to a big increase in sales. In 1896 he sold 181 cars, which was more than all the UK and USA manufacturers combined, and in 1900 sales were up to 603 cars.

Benz failed to keep pace with the rapid progress in car design. This dogcart dates from 1898, but he continued to produce roughly the same model until 1908.

But Georgano described Benz as 'almost endearingly unworldly'. He failed to keep up with the rapid technological progress that flowed from his beginnings and was still producing essentially the same model at the start of the twentieth century. Whilst it was simple and reliable, and in its early days quite advanced in its thinking (with electric ignition, rather than a hot tube) it had by then become almost laughably out of date in other respects. He refused to accept the near-universal sliding gearbox, or that the friction clutch was superior to the driving belt. Nor would he espouse the vertical engine. He also failed to provide the higher-revving engines that had become near universal. The plodding motor that had delivered a meagre 400rpm in 1886 had only been upgraded to 700rpm by the turn of the century, when the likes of De Dion Bouton were achieving more than twice that. Benz preferred his engines that way, thinking they were less likely to shake themselves to bits. Sales fell, from 603 in 1900, to 226 in 1902 and just 172 a year later, but the company continued to make them until 1908. Benz was increasingly sidelined. He resigned from the company in 1903 and, although he was to live for a further twenty-six years, was little involved in its business affairs thereafter.

GOTTLIEB DAIMLER (1834-1900)

At about the same time as the Benz was emerging, another pair of pioneers were developing their own vehicle. They lived just sixty miles from Benz but, as far as is known, they were not aware of each other, or of each other's work.

Gottlieb Daimler was the son of a baker who was first apprenticed to a firm of gunsmiths and then went on to work for a company building railway rolling stock. He saw Lenoir's

One of the last pictures of Gottlieb Daimler, taken just before his death in 1900.

gas-powered internal combustion engines at work at the Paris Exhibition and later worked for W.G. Whitworth at Coventry (the city was the early home of the British car industry, and the firm later formed part of Armstrong Whitworth). Daimler returned to Germany to run a machine shop at a Lutheran school for orphans, the Bruderhaus at Reutlingen, at which one of the pupils was one Wilhelm Maybach. They were to become life-long business partners.

The principles of the Barsanti-Matteucci motor, along with the work of Lenoir, had been further developed by a German company *Gasmotorenfabrik Deutz*, or the Deutz Gas Motor Works. Their improvements were patented in 1866. The founders of this company included the engineer Nikolaus Otto. By 1872, the company had been joined by Daimler and Maybach, as chief engineer and chief designer respectively, and the lighter and smoother-running engine they developed for the company soon began to sell in significant numbers – 2,000 in Europe by 1875.

Despite the various improvements they made, these early engines suffered from a fatal flaw. Their ignition system consisted of a slide-valve which, momentarily, exposed the fuel-air mixture in the cylinder to a naked flame. Such a system was never going to be capable of running at the sort of speed needed to power a vehicle efficiently. Benz and Maybach split from Otto and his business partner Eugen Langen in 1882 and set up on their own in a house at 13 Taubenheimstrasse, Cannstatt, a suburb of Stuttgart. They started to build small, light and (by the standards of the day) faster-running engines, capable of up to 900rpm, using a 'hot tube' ignition system (whose working is described later in the book).

Daimler was granted German Patent No. 28022 for his engine on 16 December 1883. His business was originally built on stationary applications for his engines and, whilst he had long harboured ambitions to build a self-propelled vehicle, it was not to cars that he next turned. The German Admiralty gave him an order for small boat engines and, in 1885, he also produced a crude motorcycle. This direction may have been partly to do with the limited power output of his first engines for, whilst they were dramatically lighter than the competition (around 80kg, compared with the 300kg or so of many of their competitors), their power output was only around 0.5hp, and thus only suitable for lighter applications such as these. However, by the following year he had produced his first horseless carriage. This had an up-rated 462cc single-cylinder engine capable of producing 1½hp at 700rpm and giving a top speed of 10mph.

DAIMLER THE CAR BUILDER

Unlike Benz, who designed much of his motor car from scratch, Daimler initially contented himself with fitting one of his motors to an ordinary, only slightly modified, horse-drawn carriage. He bought it, ostensibly as a present for his wife's forty-third birthday, and began adapting it. Parts of the structure were strengthened, the shafts were removed and a primitive steering gear of Daimler's own design, involving the whole front axle pivoting around a central point, was installed. The engine was located between the front and rear seats and the drive was transmitted by a system of belts turning a small gear wheel, which in turn drove a sprocket wheel attached to the inside of each rear wheel. Unlike Benz's car, which had a differential gear essentially the same as in a modern car, Daimler's turned corners by the very crude means of having a system of leather discs, which slipped as the rear wheels revolved at different speeds.

Daimler Motoren Gesellschaft was set up in 1890. Its backers wanted the company to concentrate on the popular stationary motors, but Maybach in particular kept getting drawn back into car design. They built only a handful – somewhere between twelve and nineteen cars – during the period 1893-95. But after this a new model, the *riemenwagen* or belt car, went into more sustained production, with about 150 being produced in the period 1895-97.

In France Daimler entrusted an old friend named Sarazin to manage his patents. He died shortly afterwards, but his widow so impressed Daimler with her business acumen that he allowed her to assume her late husband's role. She duly remarried and, under French law,

Emil Jellinek, diplomat, car salesman and racing driver, and his daughter Mercedes.

the patent rights she had inherited became the property of her new husband. He was Emile Levassor, a manufacturer of woodworking machinery with his partner Rene Panhard. Thus was born the motor company of Panhard-Levassor, who not only manufactured Daimler engines but also re-thought the design of the vehicles into which they were put from first principles. They were the first to opt for what was to become the near-standard arrangement of a vertical front-mounted engine driving through a clutch, gearbox, countershaft and side chains. One of their first customers was Armand Peugeot, who produced his own first car in 1891. Daimler was greatly impressed with Levassor's revolutionary approach to car design, though the same cannot be said for Levassor's reaction following a visit to Daimler's production at Cannstatt.

Levassor was one of the factors that led Maybach to radically rethink his ideas about car design. Another major influence was not an engineer at all: 'I don't want the car of today, or the one of tomorrow, I want the car of the day after tomorrow!' (Emil Jellinek, taken from Jellinek-Mercedes, page 79)

Emil Jellinek was a wealthy Austrian who held the post of Consul-General at Nice for the Austro-Hungarian Empire of the day. He was a pioneering motor racing enthusiast and noticed that the winners of the early races almost always seemed to be powered by a Daimler engine or one covered by Daimler patents, but that no actual Daimler cars took part in these events. As a result rivals such as Panhard-Levassor were getting a good detail of publicity and the commercial success that came with it. Jellinek started his relationship with the Daimler Company as the customer from Hell. Despite having no technical background in the manufacturing of cars he wrote to them demanding that they produce him a car capable of the hitherto unheard of speed of 40kph. Daimler thought this request impossible and told him so. Jellinek's reply was characteristically blunt: 'Don't rack my brains, make stronger engines. How you achieve it is your affair. I herewith order four cars and will pay for them. You don't risk anything, the 40kph is my business.' (Jellinek-Mercedes, page 68)

On receipt of these cars Jellinek was still not satisfied, and demanded that they make him a new batch, this time with four-cylinder front-mounted engines. Again Daimler demurred, saying that a front-mounted engine of that size would make the vehicle unsteerable, but Jellinek overruled his objections and ordered a further six cars to his 'impossible' specification.

This led to a new Cannstatt-Daimler model being produced from about 1897, with a front-mounted engine, four-speed gearbox and chain final drive. The standard engines for these were a 1060cc 4hp two-cylinder or an 1845cc 6hp four-cylinder engine, but wealthy clients could specify additional power for competition purposes. However, these cars had a short wheelbase, were high, top-heavy and difficult to steer. They were entirely unsuited to competition or to the top-of-the-range 5.5-litre 24hp engine that some customers demanded. These shortcomings would lead to tragedy, to the design of one of the most revolutionary cars in the history of motoring and to the emergence of one of motoring's household names.

Daimler's two-ton monster, the racing version of the Cannstatt-Daimler, was described by Nixon as:

> …One of the most dangerous, cumbersome and unwieldy motor vehicles ever produced. It… had a 28hp [actually 24hp] engine of four cylinders placed high in front, a very short wheelbase and a top speed far out of proportion to its general design. These cars ran in some of the early Continental races and were responsible for at least one fatal accident.
>
> (Nixon, page 41)

In the March 1900 La Turbie hill climb, one of them crashed, killing Wilhelm Bauer, the driver and Daimler's factory foreman. Another ended up in a ditch a few days earlier during another race, and a third suffered serious and dangerous mechanical failure when two wheels broke. Jellinek himself actually drove a Cannstatt-Daimler in a race (finishing last) which may have given him a closer appreciation of just how dangerous a beast it was.

Jellinek's experience prompted him to go back to Canstatt again and ask the company to produce something even more radically different, with a longer wheelbase, more power and a lower and more controllable design. As an incentive he asked Maybach and Daimler's son Paul (Gottlieb died of heart trouble in March 1900) how many they could hope to produce in the coming year. On being told thirty-six, he said he would take the lot and wrote out a cheque for 550,000 marks on the spot. He also undertook to become sole concessionaire for France, Belgium, America and Austro-Hungary (and in due course sold all the cars at a healthy profit).

He attached one final condition to the sale. There was still a strong prejudice in France against any car with a German name, and he therefore asked that the new model be named after his daughter, whose name Jellinek himself used as a pseudonym when racing. The use of aliases was quite common for gentleman racers in these days (Brooklands used to charge drivers an extra two guineas for the privilege of racing under an assumed name). Scott-Moncrieff suggests two possible reasons for this: one was 'to emphasise the status of the competing gentleman chauffeur' and the other was to disguise their activities from disapproving relatives, who might cut off their allowances or bribe the mechanics to render their cars unable to race. Posterity may speculate as to how the company would have fared had the daughter been christened Gertrude or Daisy-May! In fact her name was Mercedes.

After the disaster that was the Cannstatt-Daimler, the new Mercedes turned out to be one of the most revolutionary and influential designs in the history of car-making. Gottlieb's son Paul Daimler had already been working on a two-seat prototype with a number of the features Jellinek had been looking for, including a honeycomb radiator, low-tension magneto ignition

A 1903 28–32hp Mercedes. The first Mercedes cars produced by Daimler made every other model seem five years out of date.

A 7,247cc Daimler Mercedes Rennwagen (or racing car) dating from 1913, seen here newly restored at its 2008 track debut.

and a one-piece casting for the engine and gearbox. In addition, the new Mercedes engine had an unprecedented power to weight ratio of 15.2lb/hp (in part made possible by a light alloy engine block), vertical valves operated by an outer timing shaft, advanced petrol and ignition system, a lower chassis with a more steeply raked steering column, a pressed steel frame, twin jet carburettors, internal expanding brakes on the rear wheels, a water-cooled countershaft brake and an improved gearshift.

The first appearance of the new Mercedes was anything but auspicious. Design work did not start until the end of April 1900, and the specified delivery date of 15 October of the same year was never a realistic prospect. Even when the first car finally appeared at Pau on 22 November it suffered from being severely under-prepared, stripping its gears, seizing its bearings and its clutch failing. Its next appearance, at Nice in March 1901, could not have been more different. It impressed with its silence and flexibility, won the Nice–La Turbie hill climb and averaged 51.9mph over the flying kilometre on the Promenade des Anglais. As Nixon put it: 'In a night, every car in existence became five years out of date.'

DAIMLER THE ENGINE MAN

In the meantime it was the development and commercial success of Benz's engines that had been Gottlieb Daimler's prime interest. By 1889 he had produced a two-cylinder twenty-degree V engine of 565cc that produced 1.75hp at 920rpm. It also featured water cooling, with a centrifugal pump and finned radiator, and a car with it fitted called the *stahbadwagen* or steel-wheeled car, was exhibited alongside the Benz at the 1889 Paris Exhibition. Daimler engines proved very popular, and by 1890, some 1,900 licences for their manufacture were granted to third parties. Among those taking up these licences was William Steinway of New York, better known for his pianos. Not everybody was keen to take up the Daimler patents. The French gas engine manufacturers Rouart turned them down on the following dismissive (and short-sighted) terms:

> Daimler invented no new principle and contented himself with making his engine turn much more quickly than did those of his predecessors. We absolutely refused to engage in this affair, which did not seem to offer the slightest interest to serious constructors.
>
> (Karslake and Pomeroy, page 105)

Daimler and Benz were to merge in 1926.

But the age of the road-going steam vehicle was not yet over. In February 1899, one of a new generation of Serpollet steam cars was displayed in London. It was petrol-fired, and had what was called a flash boiler – one that boiled water just an ounce or two at a time. It thus had a much faster start-up time than a conventional steam vehicle and, because it only produced enough steam for its immediate needs, did not emit much smoke or vapour. It was also far easier to control its engine speed than with the governed petrol engines of the day. Had Leon Serpollet produced such an advanced model of steam car ten years earlier, before the rapid development of the internal combustion engine, some think the history of the motor car might have been very different (though, as we shall see later, even this advanced form of steam car had its drawbacks). The steam car was also no mean performer in competition. In 1902 Serpollet himself set an outright world land speed record of 120.771kph (75.043mph) in his four-cylinder steam car and, at the time of his death in 1907, another steamer, the Stanley Rocket, had pushed that record up to 121.544mph.

Leon Serpollet on his 1887 steam-powered tricycle. Serpollet made a pioneering trip from Lyon to Paris in five days in 1890.

CHAPTER 4

CARRIAGES WITHOUT HORSES

The foolish conservatism which models the horseless carriage on the same lines as the carriage drawn by horses will disappear.

(The Spectator, 1896)

We see no reason why anyone should expect a carriage propelled by a motor to differ in looks from the familiar horse-drawn model.

(The Worshipful Company of Coachbuilders and Harness Makers,
speaking to Autocar Magazine in 1897)

Horse-drawn carriage traditions dictated the form of early motor cars, often in the most inappropriate ways. Not for nothing were the earliest cars referred to as 'horseless carriages', for that was precisely what many of them were; carriages, with an engine to replace the horse. A product was actually marketed that made the direct substitution. The British entrepreneur Harry Lawson (of whom more, later) in 1899 launched the *Motorwheel*, a single-wheeled self-contained propulsion unit, one of whose applications was to be fixed between the shafts of a carriage. As Georgano shows, the bodywork of early cars evolved directly from the tradition of horse-drawn carriage-building, and many of the design features of those carriages were carried forward, regardless of whether or not they were applicable to the new mode of transport. To some the continuity of the coachbuilding tradition was seen as a virtue, and the attempts to design a functional motor car were regarded as positively vulgar by comparison. Even *Scientific American* appeared to take this view: 'The work of Mr Woods on behalf of his company has been exclusively toward the production of fine artistic carriages and all the styles and characters known to the carriage-maker's trade, rather than the mere production of a self-propelling machine.' (*Scientific American*, 13 May 1899)

Thus, the early cars were built unfeasibly high off the ground (logical in a carriage where the driver needed to see over the horse, but irrelevant in a car). Some had banks of seats facing each other, leaving the driver to peer round the passengers facing him and sitting in front of him, in an arrangement known as *vis-à-vis* (its opposite, *dos-a-dos*, had the rear passengers facing backwards, sharing the same back-rest as the forward-facing front passengers). Storage space for luggage was often provided on the roof, further adding to their high centre of gravity. A short wheelbase was held to be aesthetically pleasing in a horse-drawn vehicle, and the use of wood for the chassis, being less strong than steel and more inclined to sag, also tended to impose a shorter wheelbase. Again, this was not a problem for a horse-drawn carriage, but a short wheelbase on a car meant fidgety handling. It imposed other design constraints such as the necessity of rear rather than side entry (a problem made worse by the unnecessarily large rear wheels on most early cars). One consequence of this was that the rear passengers had to step out of their vehicle into the filthy dung-covered roads of the day rather than onto the relative cleanliness of the pavement. Drivers sometimes used to get round this by parking their vehicles sideways on the kerb.

Short chassis were also partly responsible for the early fashion of setting the rear seats very high up – a style known as the *Roi des Belges phaeton*. This was named after the King of the Belgians, after he ordered a car in the company of his mistress, who specified this feature and demonstrated what she wanted by laying out some of the elaborately upholstered easy chairs in the room they were in. She even had the manufacturer copy the style of the upholstery. One practical benefit attributed to this design was that it gave the rear passengers slightly better protection from the dust that was one of the banes of early motorists' lives.

On some early cars the front wheels were made smaller than those at the back, in the carriage-building tradition. On horse-drawn vehicles with traditional centrally pivoted steering this was necessary to allow the front wheels to pass under the frame as the coach turned, but was quite superfluous for a vehicle with Ackermann steering. This latter feature made it possible for carriages and cars to have front wheels as large as the rear ones, which also made the

The practice of storing luggage on the roof added to the top-heavy handling of early saloon cars. This is a 1900 Panhard et Levassor.

vehicle mechanically more efficient. However, this was widely resisted by the coachbuilding fraternity on no better grounds than that it was felt that equal-sized wheels 'would not look right', and this prejudice was carried forward by some of the early coachbuilders. In some respects it would even have made more sense to reverse the arrangement and have the larger (steered) wheels at the front, on the basis that holes in the road have a much greater effect on the steering of small wheels than on larger ones, and that tyre wear/damage is correspondingly increased on smaller wheels. Peugeot in 1895 were one of the first to establish near-equality in the sizes of front and rear wheels

Another feature that carried over in Britain from coaching days was a preference for open cars (97 per cent of the cars made in the United Kingdom in 1912 were open tourers, and as late as 1927 new car registrations in Britain were made up of 89,000 open-topped vehicles and 76,000 closed). The British had a tradition of open-air travel. For the fashionable urban dweller the coach was a place to see from and be seen in, making the open top *de rigeur*, whilst in rural areas open-air travel was held to have health benefits. There were also practical reasons for it. First, open vehicles were cheaper to build. Second, the weight of a fully enclosed body on an early car could be twice that of its open equivalent, and this could have a dramatic impact on the performance of these low-powered vehicles, particularly as designers were struggling to cope with their customers' demands for ever-greater luxury (and hence weight). Weight was a general problem with coach-built bodies. A touring car chassis of the mid-1900s might weigh 15cwt, but its coach-built body could add a further 7cwt. For a large limousine the body could weigh over half a ton.

Another consideration, as closed cars got ever heavier and remained rigidly mounted on the chassis, was that the poor road surfaces on which they travelled tended to shake them apart. The practice of using the roof for storing luggage was by no means helpful in this respect. Even the best coach-built body would very quickly develop annoying rattles. Open vehicles tended to have a greater degree of innate flexibility. Herbert Austin, no less, complained in a speech in 1911 of the difficulty of building a satisfactory fully closed body.

But changing fashions in relation to weather protection also reflected social change. In Europe in particular the market for early cars was squarely aimed at the wealthy classes who would hitherto have purchased horse-drawn carriages. The tradition here was that, whilst the pampered passengers enjoyed full weather protection (either in the form of a fully enclosed passenger area or a hood) the servants who drove them were, to varying degrees, exposed to the elements. The popular landaulet car design, incorporating these features, was a direct translation from the horse-drawn carriage of the same name, and was described by the *Autocar* magazine of 1906 as 'the covered motor car of the future'. Other models extended the hierarchical principle even further. The 1914 Wolseley, as bodied by Pytcheley, featured a footman's seat, fully open to the elements, mounted to the rear of the (weather-protected) passenger compartment. Happily, this feudal anachronism did not catch on.

It was the growth of a new breed of owner-drivers, after about 1908, that changed all this. Exposure to the elements was all very well for the working classes, but was not acceptable to those who owned the cars. Enclosing the driver's area in turn necessitated other changes in design. The windscreen (where fitted) tended to mist up more frequently with an enclosed car, making it necessary for the driver to open it more frequently to clear it. Because it was more convenient to open the driver's side only, this led to the fashion for V-shaped windscreens. Prior to enclosing the driver, car windows tended to be opened and closed by means of a leather strap, in the manner used on railway carriages of the period. Operating these was very inconvenient for drivers wishing to make hand signals, and led to the introduction of the sliding variety, which offered greater flexibility of opening for ventilation.

But perhaps the biggest hangover from the horse-drawn age, and one which inhibited the development of the motor car as a form of mass transport, was the adherence to traditional coachbuilding techniques with regards their bodywork. They would typically have an ash frame, into which would be set panels of, perhaps, mahogany or walnut, painstakingly finished

The 1902 8hp MMC voiturette which won a gold medal in reliability trials.

with coat after coat of hand-applied paints and varnishes (as described below). It was reckoned that, in 1913, a full man-year of labour could go into the making of a not especially luxurious coach-built body. Such processes were wholly inappropriate to a new form of mass transport. One of the most opulent examples of the coachbuilder's misapplied art was the British MMC that won a prize for the most luxurious coachwork at the 1903 Paris Show. It featured curved glass windows, lavish upholstery, folding tables, a barometer, a clock and both a speaking tube and a ship's telegraph for the owner to convey his wishes to the chauffeur.

The traditional coachwork painting method:

First the surface was cleaned and slightly roughened to help key the first coat, the priming coat of lead oxide, turpentine, and hard drying varnish for metal surfaces, or white lead, linseed oil, turpentine and lampblack for wooden parts. Then came two to six coats of filling up to make good small imperfections on the surface. These could be applied with a putty knife or with a brush. Stopping to hide screwheads, etc. then followed. Next the surfaces were smoothed by rubbing down with a wet artificial pumice stone, or waterproof abrasive papers. The surfaces were now ready for the ground colour coats; the number applied depended upon the colour. These coats were smoothed down using moistened felt pads dipped in pumice. Abrasive paper was also used and so were strips of cuttle fish bone, so the process became known as 'fishing down' as well as 'flatting'. The flatting operation was intended to remove brush marks as well as other minor irregularities. Then came a coat of glossy colour which was flatted. Finally, several coats of varnish were applied...

The paintwork had to be treated with care, it scratched easily and needed careful washing down and leathering with a chamois leather after exposure to the elements.

(Newcomb and Spurr, pages 403-04)

In pre-car days each large town or city would have had one or more coachbuilders, catering to their affluent local clienteles. Many of the best-known coachbuilders who provided the bodies for the luxury cars of the early twentieth century had histories reaching back to the early nineteenth or eighteenth centuries, if not before. Barker, who was to become for many years the 'official' body builder for Rolls-Royce, dated back to 1710, and had built horse-drawn coaches for George III, William IV and Queen Victoria. Hooper had started out as Adams and Hooper in 1805, and again had been suppliers to Queen Victoria and Edward VII. Arthur Mulliner of Northampton, who built the bodywork for Daimlers from the 1890s, had started out in 1760 specialising in mail coaches. But the daddy of British coachbuilders was the Huddersfield firm of Rippon. Their business dated back to 1555, when Walter Rippon made a coach for the Earl of Rutland, and who also numbered Queens Mary and Elizabeth I among his early customers. With this long tradition came a tendency towards conservatism and inertia, which all too often put them in conflict with many of the motor manufacturers and the motoring public.

The technology of horse-drawn carriages being fairly static and the coaches being long-lasting (due in part to the lower mileages they covered) meant a coach was, usually, a one-off purchase. This made individuality an important factor, and customers generally did not want something off the shelf. But by 1906 *Autocar* had begun to recognise the differences between the carriage and the car markets:

> It was but a couple of years ago, that when an ordinary provincial carriage builder was asked to build a body for a motor car he treated it rather as a joke... (Now) they have begun to understand that a new motor car is often bought every year by a customer who was content formerly with a new horse carriage every ten years at the most, and whose finest conveyances were probably inherited from his ancestors.

The same was not necessarily true for the more humble end of the body-building market, where the manufacturers of commercial vehicles were beginning to recognise the benefits of standardisation and mass production.

Notwithstanding these tensions, most early motor manufacturers left it to others to determine the bodywork in which their vehicles would be clad. This could lead to other inefficiencies in the manufacturing process. By the late 1890s Daimler would send his rolling chassis by rail from Coventry to Northampton where Arthur Mulliner would undertake the painstaking process, which could last months, of coachbuilding their bodies.

Nowhere was the tension between traditional coachbuilding and the demands of mass production thrown into sharper relief than in the case of the Argyll Company, one of Scotland's early car-builders. Founded in 1899 under the name of Hozier Engineering Company, a series of successes in trials and races from 1901 onwards saw the marque being spoken of as one with a future. In 1905 the company produced 1,200 cars and paid a dividend of 35 per cent to its shareholders. The company re-formed itself into Argyll Motors Ltd, with a share capital of £500,000, and promptly embarked on the construction of a huge and lavish new factory on a twenty-five-acre site at Alexandria in the Vale of Leven. The grounds included a test track and a test hill and the building itself cost the enormous sum (for the day) of £220,000. *Autocar* reported its opening in the following breathless terms:

> Nothing has been omitted that can tend to add to the comfort of the workpeople. The sanitary arrangements are above criticism, and the space devoted to lavatory and cloakroom accommodation for the workpeople occupies as much ground, and must have cost as much money, as many a factory complete.

> (*Autocar*, April 1906)

This approach was carried right through the building, with marble staircases, executive accommodation fitted out in the Georgian and Elizabethan styles, and the provision of large dining halls and lecture theatres. The factory cost £12,000 a month to operate and demanded a volume of output that necessitated modern production methods. But the company was rooted in the hand-built traditions of the past. The engines were hand-made and individually tested at length. Each chassis had a 100-mile road test and the coach-built bodies each needed thirty to thirty-five coats of paint and varnish. Even so, the company might have been saved had not its driving force, Alexander Govern, died unexpectedly at the age of thirty-eight in 1907. The company seriously lost its way. Output struggled back up to 452 cars by 1910, but the factory's running costs demanded many thousands of cars to roll off the production line each year, and it was never more than half full. The company struggled on until June 1914 when it went into liquidation, the management accused of being 'not fit to run a hen coop'. It was later re-born, a shadow of its former self, and struggled on in a small way until production ceased entirely in 1928. In many ways the factory was twenty years or more ahead of its time, but it lacked a Henry Ford to match its production methods to its other aspirations.

STYLES OF EARLY CAR BODY

Having identified the continuity between horse-drawn and motorised carriages, we can see how the main types of horse-drawn carriage were translated into cars. The names of different styles of horse-drawn coach tended to vary between one area and another, and between one manufacturer and another. But the styles of early motor car often took not just their design inspiration, but also their names, from their horse-drawn predecessors. Some of the most common were:

THE CABRIOLET: This was originally a two-wheeled one-horse design, imported from France in the 1790s. The name derives originally from the Latin word for goat, *caprioles*, and its horse-drawn predecessor was said to have capered like a goat. It carried two people and generally had a seat or platform at the rear for servants to sit or stand on. Weather protection consisted of a leather hood, a hinged leather apron that spread over your legs and (if you were lucky) a folding wood and glass screen that could be raised in front of the passengers. By 1815 it had become the most fashionable vehicle for town use. At the same time its drivers had become so notorious for their reckless driving that attempts were made to ban them from the streets. Their speeding was not just a danger to pedestrians; any servants clinging on for dear life at the back were in danger of being hurled into the street at every corner. Many of these vehicles plied for hire, the shortened form of their name giving us the popular term for a taxi. The motorised version was a soft top, usually with two doors.

THE BROUGHAM: This was named after Lord Brougham, one-time Lord Chancellor during the reign of William IV, who was said to have ordered the first example of such a vehicle in 1839. It was a small one-horse four-wheeled carriage with an enclosed passenger compartment of about 4ft front to rear. There were glass windows in the doors and front. Among its other design features was an *opera board* – an early form of bumper, consisting of a stout piece of wood that went across the back of the vehicle, above the rear axle. This was no mere ornament: in the event of a nose-to-tail collision, the pole of the horse-drawn carriage behind could well go through the passenger compartment of the vehicle in front, impaling the unfortunate occupant. This aspect of the design was, of course, entirely superfluous in the motor age. The car was a two-door coupe with covered rear seats and open front ones.

THE LANDAU: This was a four-wheeled, usually four-seat, coach driven by a servant who sat on a front box. The top half of the body would be of folding leather. The originals came from the town of Landau in Germany in about 1747. They were at first quite large and cumbersome vehicles for town use and, with the hood up, extremely dark, dirty and smelly for the passengers, since the hoods of the early models were waterproofed with grease and soot. Later versions had a much cleaner patent leather top, and the *side-light* or *five-glass* landau had the front part of the body fixed and glazed, with only the rear part opening. The motorised equivalent of this was called the motor landau or landaulette, and was the most popular form of closed car just prior to the First World War. This was a limousine with three rows of seats, the rear ones under a hood, the middle ones facing backwards under the bodywork, with the driver's compartment either open or with a roof, but no side screens.

THE TONNEAU: This was fashionable from about 1898 to 1905 and was a design derived originally from the governess cart (the name comes from the French word for a *tub* or *cask*). It had two rows of forward-facing seats, the rear ones entered via a door in the back. The rear entrance was made necessary by the fact that the short wheelbase and large rear wheels left no room for doors at the side. In addition to the problem that passengers had to enter and leave the car from the road, rather than the pavement, it also made it difficult to provide weather protection for the passengers. The version of a tonneau with side doors was called a phaeton.

THE PHAETON: A four-wheeled owner-driven open carriage of a kind originally driven by the Prince of Wales and his debauched companions in the 1790s (by the time he became George IV in 1820 he had to have an easy-entry version built to accommodate his gross size and general immobility). Some models had a rumble seat fitted to enable a groom to chaperone a lady driver (in the absence of such a seat, the groom might have to accompany the lady on horseback). The larger ones had up to three rows of seats – the people carriers of their day – and were known as hunting or Beaufort phaetons. Phaeton in Greek mythology is the son of Helios, who took his father's chariot – better known to us as the sun – for a joyride. So reckless was his driving that he nearly set fire to the earth, and Jupiter was forced, in an early and extreme form of traffic calming, to strike him down with a thunderbolt.

THE LIMOUSINE: This derives indirectly from coaching. It was originally a coarse, water-repellent woollen cloth made in Limousin, which was used to make carriage rugs and, later, awnings to protect the carriage driver. The term limousine came to be used for a car with an enclosed body for the passengers and an awning for the driver.

THE SEDAN OR SALOON CAR: This carried the driver and the passengers in a single enclosed compartment. The term probably derives from the enclosed sedan chair. One irony of the

A 1903 Wolseley tonneau. Wolseley was Britain's highest-volume car producer before the First World War (though Ford assembled more imported Model Ts at his Manchester plant).

A Mors, built more for comfort than speed. This is the 1902 Limousine.

Austerity motoring from a company later noted for more luxurious products – a 1904 Humber Olympia Tricar. This shows graphically how exposed the passenger in this design of vehicle was in the event of an accident.

Another design influenced by horse-drawn ancestors – a Daimler Wagonette from 1897.

design of many early sedans was that, in an effort to maintain the separation of the passenger from the lowly chauffeur, the rear seats were positioned as far back as possible. The result was that the passenger was sitting more or less over the back axle and getting a much less comfortable ride than his chauffeur, who was positioned nearer the centre of the compartment.

THE SHOOTING BRAKE: A saloon car with the boot replaced by a large luggage storage area, this vehicle was accessed from the rear and was much used by shooting parties. Their weight and driving position also made them very suitable for breaking in horses, but when the concept was translated from the horse-drawn to the motorised version, the *break* part of the name somehow got transformed into *brake*.

THE CHARABANC: From the French word for a carriage with rows of transverse seats, this was the original horse-drawn *omnibus*. (*Omnibus* derives from a Latin word meaning *for everybody* – except, apparently, peasants and soldiers, who were forbidden to use the vehicles licensed under the terms of a patent granted to the scientist Blaise Pascal to run a prototype bus service in Paris in 1662.) This type of vehicle made its first post-industrial appearance in Lyons, France, in 1828, and a London operator, George Shillibeer, copied the idea and started running them in the capital the following year. The idea of a double-deck omnibus came in around 1840 and, by the turn of the century, some 4,000 horse buses were running on the streets of the capital. There were still 3,500 running in 1905 when the last horse-drawn bus was built, and the last service to run, between Newcastle and Gateshead, was only discontinued in 1931.

The charabanc, more associated with excursions than with scheduled services, made its appearance in pre-car days, and its motorised form was seen before the First World War. In 1912 one of them went out of control on a hill near Durham, killing ten members of a co-operative choir. They rather quickly became a by-word for impropriety, in particular the number of raucous women who, it was alleged, were now able to inflict themselves upon holiday destinations on Bank Holidays. According to Turner:

> The Medical Officer of Health for Chelsea, Dr Louis Parkes, told the Royal Commission on Divorce, which reported in 1912: 'One frequently sees a number of births returned within about nine months after the Bank Holidays'.
>
> (Turner, page 161)

CHAPTER 5

SPEED! THE FIRST
ROAD RACES

Victories will bring world fame. The winning make will be bought and will always be bought.
If you do not enter, the conclusion will be drawn that you are unable to enter. It would be
commercial suicide to abandon racing.

(Emil Jellinek, quoted in his son's biography of him)

I never thought anything of racing, but the public refused to consider the automobile in any
light other than a fast toy. We had to race.

(Henry Ford, quoted in *Bak*, page 39)

Catching sights of little puffs of dust in the distance that betray a competitor about to be
overtaken, I have often given myself up to the intoxication of the moment...

(Marcel Renault, a few days before his death,
whilst overtaking in the Paris-Madrid race)

...irresponsible motor maniacs who slaughtered livestock, polluted the atmosphere, terrified
horses and made the day unbearable with their noisy and noisome contraptions.

(A more critical contemporary view of motor racing,
quoted in *Villard*, page 62)

THE VERY FIRST RACERS

Motor cars were very soon capable of speeds way beyond those of the road-going vehicles that
had preceded them. The desire of men – and a good number of women – to race them became
irresistible. The commercial imperative to race referred to by Emil Jellinek (above) was also
powerful. As manufacturers found out, when they took up, or dropped out of, racing, success in
competition sold cars in large numbers. The net result was a dramatic and rapid development
of the technology, which found its way very directly through to the models ordinary customers

bought. In less than a decade from the first real race to the end of great road racing era the maximum speeds of the cars taking place increased eight-fold and many aspects of car design also evolved almost beyond recognition. What follows can be no more than a brief account of some of the main events from a very large body of racing which took place over this period. The events covered range from those publicised as reliability trials to out-and-out tests of speed, not least since the rules under which they were conducted (sometimes in an effort to assuage peoples' concern about racing on public highways) can make it difficult to separate out the two.

1894: PARIS–ROUEN

What is generally acknowledged to be the first motor race took place on Sunday 22 July 1894. It was organised by the newspaper *Le Petit Journal*, which had hitherto promoted cycling events, and ran the seventy-eight miles between Paris and Rouen (there had been earlier speed trials along the banks of the Seine, in the area of the Bois de Boulogne, as early as 1887 and 1891, though for the first of these only a single entrant, Count Albert De Dion, turned up).

The 1894 Paris–Rouen was not quite a race in the sense we understand it today. The starters – thirteen with internal combustion engines and seven steamers – were to be judged not just on speed but also on whose vehicle best fitted the description 'without danger, easily handled and of low running costs'. So, strictly speaking, it might be regarded as some kind of reliability trial, though the event was subject to a time limit, qualification involved a speed trial and much of the speculation of the public revolved around who would finish first.

An improbable racing car, the De Dion steam tractor was first home in the Paris–Rouen event (not officially described as a race) in 1894.

As with other pioneering transport competitions – the Rainhill trials on the Liverpool and Manchester Railway spring to mind as a comparison – it drew every kind of eccentric out of the woodwork. Among the original 102 entrants there were cars said to be powered by: gravity, mineral oil, compressed air, a system of pendulums, hydraulic power, an electro-pneumatic vehicle, and others operated by a multiple system of levers, the weight of the passengers, and something called a self-acting mechanism. Many of these turned out to exist only on paper, or in the fevered imaginations of their inventors. None of them completed the 50km in the four hours necessary to qualify for the event and, at the start line, the competition turned out to be solely between steam- and petrol-driven vehicles.

But even the successful entrants hardly measured up to our modern conception of racing cars. The first to depart was the Count De Dion in his two-ton steam tractor, pulling a two-wheeled landau which carried four people sitting *vis-à-vis*, and the rear of the field was brought up by two Serpollet vans, seating eight and ten people respectively. The anticipated pace of the event may be judged from the fact that the newspaper proprietor James Gordon Bennett assigned a reporter on a bicycle to cover the full length of the race. Nonetheless, the event certainly captured the public interest, with crowds lining much of the route.

The first to Rouen was De Dion, averaging 11.6mph, and one of the Serpollets brought up the rear at 6.1mph. But De Dion was not declared the winner. Under the arcane scoring system employed by the organisers his vehicle was ruled to be too expensive to buy and run (requiring a mechanic/stoker as well as a driver) and too complicated to control. He was placed behind the equal winners; petrol cars built by Panhard-Levassor and Peugeot. The leading Peugeot had finished ahead of the Panhard on the road, but the Panhard got extra points for having manufactured the engines for both vehicles. Fourth, after De Dion, was one of the Serpollet steamers, which apparently won extra points for its potential as an omnibus. De Dion went on to found the Automobile Club de France in 1895, which was responsible for organising no less than thirty-four inter-city races over the following eight years.

But the historic claims of this event as a first cannot be allowed to pass without noting that Britain has a much less-well-known claim to have staged the first race between self-propelled road-going vehicles, and this despite the Locomotives Act of 1865 setting a 4mph speed limit. It took place on 30 August 1867 between two steam carriages, starting at Ashton-under-Lyne near Manchester and finishing at the Old Trafford showground. The race started at 4.30 in the morning, no doubt to avoid the attention of the law, and the winner completed the eight-mile journey in less than one hour (at the time, a highly illegal speed). Both competitors later gave a display at the showground.

1895: PARIS–BORDEAUX–PARIS

1895 saw one of the first epic long-distance road races, covering 732 miles from Paris to Bordeaux and back, over one of the roads built by Napoleon to facilitate troop movements. This race was part sponsored by James Gordon Bennett (of whom more, later) and had prize money totalling 76,000 gold francs, half of it going to the winner (compared with just 5,000 francs for the victor of the previous year's Paris–Rouen race). It was run over three days in June, and attracted forty-six entries, twenty-three of them petrol-engined, thirteen steamers, eight powered tricycles and bicycles and two electric vehicles. First to finish, at an average speed of 15mph, was Emile Levassor, driving one of his 4hp two-cylinder Panhard-Levassors. However, the organisers, fearful of adverse public opinion, particularly if there were accidents, sought to play down the speed aspect of the contest and to present it more as another reliability trial.

To reinforce this they even ruled that cars specially built for racing (rather imprecisely defined for their purposes as two-seaters) could not win the race. This denied Levassor the prize money, even though if ever a man had earned it, it was he. His drive was an epic of endurance. He was due to pick up a relief driver at Ruffec, after having driven for fifteen hours since the start. However, when he arrived there, well ahead of schedule at 3.30 in the morning, the relief driver was still fast asleep in bed. Rather than wait to rouse the man, Levassor simply carried on. By the time he pulled into Bordeaux at 10.40 that morning, he had been at the controls of this primitive and demanding vehicle for almost twenty-three hours (among its other features, it had tiller steering, iron-shod wooden wheels and candles for its only lighting, which cannot have made it the most relaxing of vehicles in which to drive through the night).

Without further ado he signed himself in, drank a glass of champagne, turned around and immediately began the return journey. He was not to know that his nearest rival was already some four hours behind him. When he arrived back at Ruffec, the relief driver was ready and waiting, but Levassor was by now just getting into his stride and refused to let him drive. By the time they reached Paris, at 12.57 p.m., Levassor had been driving for forty-eight hours and forty-seven minutes, with just one break of twenty-two minutes (over and above the fuel stops every 100km). He finished six hours ahead of the next competitor, and over forty-one hours ahead of the only steam vehicle to finish the race. A monument to his feat was erected at the finishing point in 1907. At the celebratory event to mark the end of the event someone was bold enough to predict that average speeds of over 20mph would soon be attainable, but Levassor was quick to downplay any such wild speculation, saying: 'There's always someone at this sort of banquet who makes some ridiculous remark.'

The possibility was that Levassor was trying to allay the fears of nervous authorities at the growing speed and popularity of the motor car, something that would very quickly become a real issue for the industry. The identical twin to Levassor's 'winning' car was promptly bought by the would-be English car magnate, Harry J. Lawson (whose dubious career in the British motor industry is documented later in this book).

One of the Peugeots in the race was driven by a man whose name would become famous in motoring circles: Edouard Michelin. As we see later, Dunlop in effect re-patented the pneumatic tyre in 1888, but it was the Michelin brothers who brought out an improved version in 1891. Improved, but not perfected, for in the course of the race they suffered a total of fifty punctures. Their car was known as *Lightning* – not for its turn of speed, but for a fault (some said in its steering, others because it lacked a differential) that caused it to zigzag along the road, rather like the popular representation of a bolt of lightning.

1896: PARIS–MARSEILLES–PARIS

> The destruction of animals (is a good reason) for the wearing of a leather uniform when travelling at high speed. When struck, the animal is apt to scatter very unpleasantly. Not to go into detailed description, I have seen a racing car come in from a long contest fairly plastered with feathers.
>
> (Leading, but un-named, French racing driver of the period,
> quoted in *Villard*, page 20)

This was the first of the races to be organised by the fledgling Automobile Club de France, who had an observer in each car. It was sponsored by the motoring magazine *La France Automobile* and covered a mammoth course of 1,162.5 miles. But this time it was to be covered in fixed stages, with stopping points at the end of each day – no more marathon stints or night driving for the contestants. The cars were also divided into classes: there were three wagonettes, four three-wheeled Bollee tricars, five De Dion tricycles and twenty standard touring cars.

The race can safely be described as eventful, setting standards for carnage that its successors had to work hard to beat. A brief catalogue will be sufficient to illustrate the chaos. A driver named Ferte ran down and killed a pedestrian before they were out of Paris; a De Dion steamer, way out in the lead, suffered the total collapse of its pneumatic tyres, and other competitors had more mundane mechanical failures. But, in all, twenty-seven of the thirty-two competitors made it to the end of the first stage, making for a fairly modest rate of attrition.

But overnight violent storms devastated the whole of France. Telegraph poles and trees were torn down everywhere, and the competitors had to contend with winds of near hurricane force the following day. One major challenge for the competitors, many of whom relied on tube ignition, was preventing the burners heating the tubes from blowing out. Once they were extinguished, the car ground to a standstill. This race was important in highlighting the shortcomings of tube ignition and bringing forward the development of electrical alternatives.

Fallen trees were also a major hazard that day and the fortunes of the competitors who encountered them varied considerably. Delahaye was relatively fortunate. He came upon a large fallen tree and only lost two hours in sawing it into three pieces and removing the middle one to provide a way through. Amedee Bollee had one fall directly in front of him and could not avoid driving into it, hurling his passengers out and bending the tricar into an S-shape. Another driver, Rossle, had an even more unfortunate encounter with a tree. His passengers had just helped to push their car up a hill and, on reaching the summit, collapsed, exhausted onto the road. To their horror, they saw a gust of wind catch the car and send it rolling back down the hill they had just toiled up. They ran after it in vain, only to see it written off against a tree at the bottom.

Equally exhausted were the drivers of the cycle cars. With their tiny engines of 1hp or less, and no gears, they were quite unequal to the wind, and the drivers had to pedal furiously for much of the length of the ninety-three-mile stage to maintain progress. One tricycle rider actually ran out of petrol and was forced to rely on pedal power alone for the last ten miles.

An 1897 Bollee voiturette of the kind that was first home in the Emancipation Run and took part in the 1896 Paris–Marseilles race.

Selwyn Edge's 80hp Napier passes between the crowds during the 1904 Gordon Bennett trials.

Wildlife also played its part in the devastation. Camille Bollee's tricar was eliminated from the race when it was badly damaged by a charging bull, and it was alleged that no less than fifteen dogs were run down by competitors. Smaller beasts, such as the chickens referred to in an earlier quote, did not even warrant a head-count. One of the larger dogs was that hit by Levassor, the victor of the previous year's Paris–Bordeaux race. At the time there appeared to be little serious damage to Levassor or his car, and they continued to the end of the stage. But, about a year after the race, he collapsed and died at the age of just fifty-three. He was found to have suffered unsuspected internal injuries during the accident. Despite all this, fifteen of the thirty-two starters made it to the halfway point and one of the motor-tricyclists actually held the lead, at an average speed of just over 15mph. The later stages turned into a battle between the Panhards and the Peugeots, with Mayade and Merkel coming first and second in four-cylinder 8hp Panhards.

As a footnote to this race, Merkel's car was imported to Britain and eventually came into the ownership of the motor dealer and racing driver Selwyn Francis Edge. He wanted to carry out a series of modifications to it (such as replacing the tiller with a steering wheel, solid tyres with pneumatic, and the fitting of a radiator). Through a friend he was introduced to the firm of Montague Napier, coin-weighing apparatus manufacturers, who agreed to do the work. They did so highly successfully, going on to build their own modified engine for it. Edge was so impressed that he got them to go into production, and set up the Motor Vehicle Company to sell their entire output. So was born the distinguished car and aeroplane engine manufacturer Napier.

The car still being the plaything of the wealthy, it was decided to set up a series of races in January 1897 in Nice, where the rich spent the winter season. In these races, and in others that took place that year, the four-wheeled petrol car did not perform well, being beaten – always on reliability and sometimes on speed – by the steamers, and consistently by the Bollee brothers in their racing tricycles.

The Brasier Company started producing cars (initially modelled on the Benz) in 1897. Henri Brasier was a talented designer who also worked with Mors. He formed a partnership with car manufacturer Georges Richard, and their Richard-Brasier cars enjoyed victories in the Gordon Bennett Trophy races of 1904 and 1905.

1898 was the year in which Belgium and Switzerland held their first motor races, and Emile Mors, who would be a dominant figure in motor racing for the next decade, first emerged. His company had an unusual background for a car builder, starting out by manufacturing artificial flowers. They had a machine that wrapped their wire stems in paper and this led them to adapt it for wrapping insulation around electric wires. This in turn led Mors into car production in 1896 and, unsurprisingly, at a time when most French makers were using hot tube ignition, the Mors cars opted for the electric variety. Like Bugatti after him, Mors produced racing cars for amateur drivers. He also organised small races for newcomers to the sport, called *les petits coureurs qui poussent*. Until legislation was passed in 1900 there was nothing in France to stop anyone staging a motor race on public roads, provided no one objected beforehand. Mors also introduced the novel idea of classifying the cars in the race by their catalogue prices.

Despite having Henri Brasier as their chief engineer, and despite considerable racing success, the company was badly managed and thrived only briefly. After booming dividends in 1902 and 1903, they sank into serious loss. Some degree of recovery was achieved under Chairman Andre Citroen from 1908. But the company never again approached the glory of its early days and was bought out by Citroen (by then a leading figure in volume car manufacture) in 1925.

1898: PARIS–AMSTERDAM–PARIS

Even by the chaotic standards of early road racing, this event – the first to cross national boundaries – holds a special place in the record books of motoring lunacy. The starting point of the race was only fixed hours before it was due to begin; the finishing point was only settled as the cars approached it, and much that happened before and during the race had a similar element of farce to it. The 889-mile race was divided into six sections, with a day's rest in Amsterdam. Sixty-nine entries were received for the racing car section alone (there was also a non-competitive touring car class).

By now races were organised to a considerable degree. The Automobile Club de France produced a handbook of rules and information that, for this race, ran to 115 pages. One item they forgot was that the start of the route was in the Department of the Seine, and that there existed in that Department an 1893 by-law requiring every car used on the roads to be approved by the authorities. In the true Gallic spirit, this law had been universally ignored by the locals and almost totally forgotten until, on the afternoon before the event, the police told the organisers that a Monsieur Bochet, the district engineer, would be round in the morning to carry out the inspection. This struck terror into the hearts of all involved, for Bochet was notorious as an autocrat beside whom Bismarck looked indecisive, and also a rabid car-hater. The following day, after a good deal of provocative messing around of the competitors, Bochet started rejecting the cars one after another, summarily and without explanation. The competitors were by now so incensed that they told Bochet that the race would start whether he liked it or not.

This provoked Bochet into some radical traffic management measures. He called out a company of infantry and half a squadron of the 24th Hussars, the latter setting up camp at the proposed starting point. Field guns were also placed along the proposed route, a rather extreme predecessor to the speed camera. The competitors responded by deciding to change their starting point to the nearest place beyond the jurisdiction of Bochet, and all the available horses in the area were hired to tow the forty-eight starters to Villiers-sur-Marne. They then realised that they had come without their petrol, and Amedee Bollee bravely volunteered to return to their garage with a horse and cart and collect the petrol from under the noses of the waiting army.

The race finally got under way, with its usual crop of mechanical breakdowns, crashes and other incidents, but the organisers were still left with the problem that the planned finish point was also on Bochet territory, and he was by now apoplectic with rage. A triumphant arrival at Versailles had been planned, and those few entrants whose cars had survived Bochet's severe scrutiny did not want to be denied their moment of glory. But the others had no taste for whatever Bochet had in mind for them, and an alternative finishing point at Montgeron was identified. In the end the triumphal arch at Versailles was taken down, but half of the officials still turned up there, the other half going to Montgeron, and many of the competitors remained confused as to whether or not they had finished. For what it was worth Fernand Charron in an 8hp four-cylinder Panhard was declared the winner, at an average speed of 26.9mph.

1899: TOUR DE FRANCE

This race, with seven stages totalling 1,350 miles, was the most ambitious yet. But by the standards of its predecessor it was almost dull in its orderliness. Alongside the usual crop of breakdowns and crashes, spectators had the sight of Girardot completing a stage with a cartwheel on one front wheel, after one of his road wheels broke, and Charron completing the last twenty-five miles of the stage to Alencon backwards, after a gearbox failure. Panhards swept all before them, taking six of the first seven places, including the winner De Knyff, who averaged just over 30mph.

In a rather more unusual development the same year the 104-mile Paris–Trouville race was open not only to motorcyclists and racing cars but also to horse-riders, cyclists and pedestrians. The contestants were handicapped and the start staggered accordingly, with the pedestrians, at one extreme, starting seventeen hours ahead of the racing cars. The idea was that, if the handicapping were perfect, they would all converge on the finish together. For anyone who is interested, it was won by a horse whose name appears not to have been recorded for posterity.

As we have seen, it was during this year that the domination of Panhard-Levassor and Peugeot began to be challenged by a new team. Mors was a French company who, in addition to artificial flowers, manufactured electrical instruments and telegraph equipment. Their first experience of racing had been the 1896 Paris–Bordeaux race where their car, modelled on the Benz, crashed into a farm cart. But by the turn of the century they were starting to be a contender, and actually won races between Paris and St Malo and Bordeaux and Biarritz. Top drivers began to switch to the car, especially after a new model, designed by Henri Brasier and more closely modelled on the Panhard, was introduced.

But the speeds of the cars continued to increase, rather more quickly than their ability to steer or stop. The race at Pau in the south-west of France in 1900 saw the winner record an average speed of 43.8mph over 209.5 miles. Serious accidents started to become much more of a feature of these events. First there was the accident at the La Turbie hill climb of 30 March 1900, referred to earlier, when Wilhelm Bauer was killed in the works Daimler. Then, during a different race for tricars, two of the contestants collided and flew into a crowd of spectators. Several were injured, including the wife of the Deputy of the Department of the Seine. First the local authorities banned racing in that area, then the Ministry of the Interior extended the ban to the whole of France. But protests were made that such a ban would stifle the country's newest engineering industry, one in which she was at that time a world-beater. These protests prevailed and racing was allowed to resume.

1900-1905: THE GORDON BENNETT INTERNATIONAL TROPHY RACES

In 1899 Mr Gordon Bennett gave a work of art to be raced for by motor cars and held as an international trophy. This is generally known as the 'Gordon Bennett Cup'. As a matter of fact, it is not a cup at all, but a model of a motor car carrying two figures in anything but motor car costume…

(*The Times*, 30 January 1903)

The French driver especially is a most reckless and daredevil sort of a fellow. He opens his throttle regardless of conditions, and does not consider for a moment that the motor is entitled to the slightest bit of consideration.

(From an address to the Chicago Motor Club by C.Y. Knight, 28 December 1909)

The name of James Gordon Bennett is inseparably linked with the early history of motor racing. The races he promoted introduced a new, international dimension to the sport. At their height the races were the biggest motor racing event in Europe, commonly referred to as 'the motor derby'.

Gordon Bennett was the son of the founder of the *New York Herald* and part-owner of the Bennett-Mackay transatlantic telephone cable. He was an adventurer who also encouraged adventure in others: he raced his yacht *Dauntless* across the Atlantic in 1870 and hired the explorer Henry Morton Stanley to go out into darkest Africa and find Dr Livingstone for his newspaper. Charming and scandalous in equal proportions, he so outraged polite society in his native America (not least by urinating in his prospective in-laws' fireplace while drunk) that he found it necessary to exile himself to Paris. There he founded the European version of the *Herald* and continued to lead a less than conventional life (for example, by racing a *fiacre* up the Champs-Elysees by night, sitting on the coachman's box entirely naked).

Although he was not personally interested in motoring (he never learned to drive, nor attended any of the races that bore his name) he recognised the newsworthiness of motor racing early on. He had helped to fund the first Paris–Bordeaux race, but had grown tired of the French domination of road racing. In 1899 he therefore put up the trophy described above (costing 12,000FF, weighing 17kg and 'preposterous in appearance', according to Jellinek) and invited the various national automobile clubs to put forward teams to compete for it. Every part of the competing car was supposed to be made in the country of the team, in the hope of providing a stimulus to national components industries. The strictness with which this latter rule was applied was illustrated when the German team tried to use French Michelin tyres, manufactured under licence in Germany. This was ruled out when it emerged that the tyres' valves were French imports. The cars also had to be of a minimum weight of 400kg, which ruled out light tricars like the Bollees.

The Automobile Club de France (ACF) were given the job of organising the first race of at least 500km, which they then named after its sponsor, rather than following his wish for it to be called the International Trophy. One of the Trophy's rules was that the country winning it one year had the duty to stage it the next. This was to present difficulties in years to come.

The first race, in 1900, turned out to be a complete shambles: the French selectors chose their team from the wealth of skilled drivers available to it by ballot, leading some of their leading motorists to resign their association with the ACF, some even registering with the opposing Belgian team (the drivers evidently did not have to be home-grown); the chosen course was found to be 60km shorter than the minimum length required by the regulations and had to be changed; there was a dispute with the French authorities about the very principle of holding the race; there were disputes within the club; the French customs impounded one of the Belgian entrants and the starting date was repeatedly changed.

As for the entrants, there were doubts as to whether the Belgian entry, a French Bolide built under licence in Belgium, really counted as being of Belgian manufacture, and the American Winton, with its single-cylinder rear engine and tiller steering, was generally mocked as being obsolete. Then there was public hostility to the principle of holding the race, following on from the two well-publicised recent racing accidents. Two days before the original starting date nobody knew when the race was due to take place or whether it would take place at all.

The 1906 Grand Prix, run at the Le Mans circuit. De la Touloubre in his Bayard–Clement adds dogs and cyclists to the hazards a Grand Prix driver had to face.

The race, when it did take place, lived up to the standard of its preparation. Only five competitors started it and only two finished. The route was un-signposted and the first part of the race was run in the darkness; most if not all of the contestants suffered accidents or mechanical failure. Perhaps the most extraordinary event in the race was when the Frenchman Charron struck a St Bernard dog at an estimated 60mph. The dog got jammed in the wheels and the car left the road. Normally Charron and his mechanic Fournier would have been killed outright; but the car leapt a ditch, narrowly missed two trees and came to rest in a field, facing the way it had come. By some miracle both car and passengers were unharmed. They were able to remove the dog (which was definitely not unharmed), re-start the car and go on to win the race at an average speed of 38.6mph. They finished with Fournier clinging on to the side of the car with one hand as he leaned out to hold a broken water pump in place. As a public spectacle, the race was a complete flop, with less than 100 spectators turning out for the finish.

THE 1900 1,000-MILE TRIAL

It is impossible to travel in France without being struck by the number of motor cars and cycles, the rapidity of their movements, and the position which, to judge from the comic and other papers, they occupy in the daily life of the community. The prominence of the motor car is due, of course, primarily to the extent and excellence of the French roads. But the steady improvement in manufacture and increase in popularity are to be attributed very largely to the series of road races organised since 1894 by leading French newspapers and by the Automobile Club de France…

Races of this kind, however, are obviously impossible in England. In the first place, the legal limit of speed is 12mph; in the second place, even if there were no such limit, there is no room. It is not merely a question of road length; the population is so dense, comparatively speaking, in most parts of England that the passage of motor cars at high speed would seriously interfere with the amenities of country life…there is a very large proportion of persons, old-fashioned perhaps, but not unworthy of consideration, who are more than content with the results already achieved and who think that no advantage is to be gained by further progress in the use of motor vehicles in England.

(*The Times,* 21 April 1900)

The principal object of the organisers of this trial was to prove…that the motor car is, even in its present state of development, a serious and trustworthy means of locomotion: not a toy, dangerous and troublesome alike to the public and its owner, but a vehicle under as perfect control as a Bath chair, capable of accomplishing long journeys in all weathers and over every kind of road with ease and safety, destined to take its place with the train and the bicycle as a common object of daily life, and as superior to them, in many respects, as they are superior to the horse and cart.

(*The Times,* 14 May 1900)

After their late introduction to the motor age, the British public in 1900 were still far from convinced about the practicality of the car as a form of transport. In some peoples' eyes too many of them were to be seen broken down at the roadside for them to be anything more than the playthings of rich (and patient) eccentrics. Some of the more remote parts of the country had barely seen a motor car. The first motor show, held in 1899 in Richmond, had hardly been an unalloyed triumph for the car. Attempts to race a horse, to prove the superiority of internal combustion, ended with the horse winning the first race by half a lap.

A group of enthusiasts led by the future Lord Northcliffe, Alfred Harmsworth, decided to demonstrate their practicality as a mode of transport by staging a long-distance trial. The route of just over 1,000 miles was the longest ever undertaken by motor vehicles in Britain up to that time. It ran from London, through Bristol, Manchester, Edinburgh, Newcastle, Leeds, Sheffield and Lincoln, then back to London, and included hill climbing at Shap Fell and three other venues and a speed trial at Welbeck Park. Given the absence of petrol stations around the country, the organisers had to arrange for supplies of petrol to be left at strategic points along the route.

The organisers made it categorically clear that it was not a race and strictly prohibited the breaking of the 12mph speed limit (8mph in towns and 10mph in Scotland). However, the target was for the cars to make the maximum speed limit their average speed for the journey, which rather suggested that speed limits were going to be broken somewhere.

Eighty-three vehicles entered for the trial and sixty-five turned up in London at 7.00 a.m. on the morning of 23 April 1900. This was a larger field than for any of the preceding French events, where emphasis on speed tended to exclude many more modest models. Many of the supposedly British entrants were indeed of a more mixed, not to say doubtful, pedigree. Harry Lawson's Motor Manufacturing Company MMCs were either very closely modelled on Panhards, or actual Panhards with MMC badges; the Triumphs were most certainly re-badged Renaults; the English Daimlers, whilst built in England, were also closely modelled on Panhards; one of the Napiers was also a re-modelled Panhard with a Napier engine; the Star, the International and the Marshall were all modelled on the Benz, whilst the New Orleans was a Belgian Vivinus built under licence. Only the Lanchesters and the little 3hp Wolseley could claim unalloyed British origins. Whilst the Wolseley fairly romped around the trial (insofar as one of its limited power could) one of the Lanchesters had to retire and the other, painfully slow and repeatedly breaking down, arrived back in London late.

The spectators who gathered at the start included many opponents of the motor car, who thought the trial would expose their fatal weaknesses and lead to an orderly return to the days of the horse and carriage. But it would be they who were confounded. The outcome of the trial was to stimulate the growth of the British car industry and it was also instrumental in the future King Edward VII ordering his first motor car. Exhibitions of the competitors were held along the route, the proceeds from which went to the Transvaal War Fund.

The trial certainly had its casualties. One of the competitors hit tramlines in Bath and did a couple of somersaults. One of the De Dion voiturettes skidded so violently that it snapped its front wheels off. Others crashed into walls, cattle or horses. But the competitors displayed their accustomed ingenuity in keeping their vehicles going – none more so than M.G. White who, somewhere between Edinburgh and Newcastle, broke the steering on his Daimler. However, he discovered that by standing on the offside running board on one leg, he could kick the offside front wheel to make it point in approximately the direction he wished to go. He completed the remaining fifty or so miles into Newcastle by this means, wearing a hole in his boot in the process. The International suffered a breakdown of its driver, being delayed by Mr Capellan bursting a blood vessel.

The speed trial on private roads was their one opportunity to show what their vehicles could really do. The Hon. C.S. Rolls won this stage, his 12hp Panhard achieving an average speed of 37.63mph.

Of the sixty-five cars that started the trial, thirty-five completed it, with twelve of them achieving an average speed of 12mph in England and 10mph in Scotland. As Nixon, who took part in the trial as a young boy passenger, said:

It registered a marked advance in the cause of British automobilism, as it decided a great many important points in design about which the early British designers were undecided. It served to settle such questions as to the choice between tube or electric ignition, wheel or tiller steering, air or water cooling, solid or pneumatic tyres, gear and chain or belt drive, twin- or four-cylinder engines for larger cars and similar problems which a severe test such as this one alone could decide.

(Nixon, page 194)

1901: PARIS–BERLIN

Things got more complicated when the races took on an international dimension. The Paris–Berlin race was an early case in point. Memories of Germany's 1871 invasion of France were still fresh and raw in many French minds, despite the fact that the French drivers would be going in the opposite direction to their military compatriots. The leading French driver Leon Serpollet was sent to reconnoitre the course and come back with reassurances for his countrymen. He told of the enormous lengths to which the Germans were going to make them welcome. The biggest problem the competitors would face, he suggested, would be the dust. There was a huge interest in the race, with 172 drivers entering, 110 actually starting and forty-seven finishing. Because of growing concern at the ever-increasing speed of the cars, a weight limit of 1,000kg (about one ton) was imposed on the entrants in the hope of making the entrants slower. Huge crowds flocked to the start and it proved difficult to clear a path for the first cars to depart.

The same was true throughout the race. Spectators, with no conception of the speed at which these cars were now travelling, would stand in the middle of the road until the last moment to watch them coming, then rush out again to see them disappear. At Montchenot, near Reims, Brasier's Mors hit and killed a young boy. Other contestants hit trams, private cars and did assorted damage to the countryside, fortunately without further loss of life.

Sections of the press railed against the event:

> With the guilty support of the public, a veritable act of human folly has been perpetrated since
> yesterday morning on the main roads between Paris and Berlin…at this moment seventy-one

Selwyn Edge in his Napier experiences intimate contact with the crowd at the start of the 1903 Gordon Bennett race.

dangerous madmen are driving over open country at speeds of express trains. These maniacs who crouch over the wheels of their mechanical carriages at 50mph knock down human beings, cyclists, cattle, anything in their path…

(*La Petite Republicaine*, 28 June 1901)

Some saw it in more class-based terms, with the working classes being put at risk for the amusement of the wealthy. In other respects the race proved to be a triumph of international relations as the Germans pulled out all the stops to ensure the competitors were plied with champagne and cigars, banqueted to within an inch of their lives and generally overwhelmed with kindness. Cards were handed out at the frontier, saying 'Soyez les bienvenus en Allemagne' (Welcome to Germany), and fore-warning them of any potential danger points they would encounter en route to Berlin. Bands played the *Marseillaise* and celebratory bombs were set off (no doubt further infuriating those who objected to the disturbance caused by the event). To add to Gallic joy the race was won by Henri Fournier in a 60hp Mors, in front of 20,000 spectators. Louis Renault won the voiturette class and the best-placed German car could only manage eighteenth. But victory was not enough to quell the growing opposition in France. On the day the race ended the Minister of the Interior announced that there would be no more road racing in France.

The ban killed off the rest of the season's racing programme in France. But the following year the ban was modified by a piece of Gallic pragmatism. The French Ministry of Agriculture had a huge surplus of alcohol made from sugar beet to dispose of each year. After a good deal of governmental in-fighting, it was decided to exempt the 1902 Circuit du Nord race from the ban on condition that all the cars ran on this alcohol fuel only. It was hoped that this would provide an alternative outlet for their alcohol lake. However, the *vin* proved to be very *ordinaire* when it came to powering cars – steam vehicles ran particularly badly on it – and the attendant publicity did nothing to popularise alcohol fuels. In addition it was said that the acrid exhaust fumes made the drivers feel nauseous. To add to their joys the race was dogged by heavy rain, making the course treacherously slippery and causing many mechanical failures. On the safety front there was good news and bad news. The good news was that only one pedestrian was sent flying by a racing car, that he was not badly hurt by it and apparently took it very philosophically. The bad news was that the injured party was the local Chief of Police, who had been trying to hold the crowds back.

The second Gordon Bennett Trophy race, also held in 1901, was possibly an even greater shambles than the first, with only one finisher, Girardot in a 40hp Panhard. A 50hp Napier had been entered, but was unable to find any British-made tyres capable of handling the stresses of an engine of that power. It had to fit French tyres and so was excluded under the Gordon Bennett 'one nationality of manufacture' rule. Because of the increasing difficulty of securing permission from the authorities to stage road races, and given the difficulty the first Gordon Bennett race had had attracting spectators, this year's race was run in conjunction with another road race from Paris to Bordeaux. Although Girardot took the Gordon Bennett Trophy at an average speed of 37mph, he could only manage ninth overall in the Paris–Bordeaux race, which was won by Henri Fournier in a Mors at an average of 53mph, followed by five Panhards. So France took the Gordon Bennett Trophy and was thus required to host the following year's event. As an illustration of the ad hoc approach to engineering at the time, Baras finished second in the light car class in his Darracq, with the engine tied on to the chassis with a length of rope, the engine mountings having broken.

1902: PARIS–VIENNA

Having at first pilloried the Paris–Berlin race, the French press were then swept away on a tide of patriotism following Fournier's victory, and lobbied for the reinstatement of racing until the

Government gave in. This made possible the next great international race, the Paris–Vienna of 1902. The problems associated with international races continued to dog the organisers of such events. The Automobile Club of Nice's attempt to run a 500-mile race to Abazzia in Italy in that same year had to secure the agreement of the French, Austrian, Swiss and Italian authorities. The route was changed over a dozen times to address objections raised by one or other of the authorities, until the Italians pulled the plug on the event the day before it was due to start.

Similar problems dogged the Paris–Vienna race of the same year, except that this time the agreement of all the governments concerned was secured, only for the French Government to lose a general election, making it necessary to re-negotiate with the new administration. Permission was finally obtained, and no less than 219 cars entered the race, 137 of them making it to the start line. The 1,000kg weight limit was once again applied to the heavy car class in an attempt to contain their growing speed, but designers were by now getting adept at combining lightness and power. As *The Car Illustrated* put it: 'The chassis of these racing cars is now a mere skeleton, but the engine is a raging lion, immense in size and power although exceedingly light.' (Villard, page 78)

The first stage graphically illustrated how much progress had been made. For the first fifty miles Fournier's Mors averaged an unprecedented 71mph, until a broken gearbox finished his race.

Improvisation was to be the key to success for many of the entrants. Charles Jarrott nursed his Panhard to the end of one stage with his wooden chassis broken. His attempts to find suitable materials to fix it proved unavailing. However, in the depths of the night sawing noises could be heard coming from his hotel bedroom. Jarrott walked rather stiff-legged from the hotel the following morning and was somehow able to fix his car and continue on the next stage of the race. The hotel staff later found that the table in his room had mysteriously lost its legs. Jarrott's improvised repair held the car together until it suffered a breakdown beyond any carpentry skills (when his car deposited the contents of its gearbox on the road five miles outside Vienna). Later in the race Louis Renault rammed a level crossing gate. He summoned the local blacksmith to help him untwist the front axle, and then he bound up a cracked differential with wire and rope and thickened the oil in his crankcase with sawdust, to stop it leaking out. Then, after being rammed and losing several of the wooden spokes in one of his wheels, he fashioned replacements from the cross bars of a chair.

The 1902 Gordon Bennett Trophy race was once again combined with this bigger event, though the Gordon Bennett entrants only went as far as Innsbruck. The epic struggle of this race within a race is worthy of being recorded separately. On this occasion the only nation to challenge the French was Britain, who put up a Napier driven by S.F. Edge and a Wolseley with Herbert Austin at the wheel.

Built on a shoestring, Edge's Napier was only ready for road trials a week before the start of the race (part of which time would be taken up travelling to Paris for the start). The team had to work all night preparing the car to leave for Folkestone (and the ferry) before dawn. As they neared Folkestone one of the cylinder heads cracked; Napier himself (who was travelling as Edge's passenger) was all for giving up, thinking that no replacement cylinder heads existed. However, Edge managed to get a long-distance phone call through to the Napier workshop (itself a major achievement in those early days of telephony) where it was confirmed that there was a spare. Two mechanics were sent with it by train to Folkestone, and fitted it during the ferry crossing. As they continued on their way to Paris the second speed of the three-speed gearbox failed. They were just able to limp it, using first and third gears only, into Paris where Napier spent a second sleepless night re-building the gearbox. They made it to the weigh-in and scrutineering with minutes to spare.

However, on the way to the weigh-in they discovered that the reverse gear had failed (and possession of such a gear was one of the requirements of the Ministry testers). A piece of bluff by Edge meant that they managed to move the car backwards a few inches, far enough to persuade the scrutineer that the car's reverse gear was in fact functioning. Then Napier discovered that, in his haste, he had left a key component out of the gearbox, without which it would never complete the race. A further sleepless night was passed rebuilding it yet again. By this time they were almost beyond caring when a misfire developed en route to the starting line. However, this mysteriously repaired itself on the actual starting line, for reasons which they never understood.

The race started and, for a change, the car ran perfectly for 200 miles, until suffering a flat tyre. At this point they found that their pump was not working. A further piece of luck then appeared in the form of fellow racing driver, Count Eliot Zborowski, who gave them his pump. (The name Zborowski is most famed in motoring circles for the monstrous aero-engined cars built in the 1920s and raced at Brooklands by Eliot's son Louis, one of which he christened Chitty Chitty Bang Bang – the inspiration for the Ian Fleming story). Further tyre problems arose, including an officious railway station employee who for some reason impounded their supply of spares. This required Edge to engage the man in a heated debate (made all the more heated by the fact that neither spoke the other's language) distracting the railwayman sufficiently for his team to be able to 'steal' back their own property.

Their next trial was a terrifying crossing of the Arlsberg Pass in Austria. It would prove to be the toughest test yet for these primitive vehicles, taking them to an altitude of 5,912ft, up – and more particularly down – hills of up to 1:7, edged with vertical drops, punctuated by hairpin bends, driving over ill-surfaced roads with no crash barriers. The roads through this pass were described in those days as:

> Hardly more than tracks, winding along precipices with nothing but boundary stones between the road and the drop beyond, crossing torrents on improvised bridges made of a few planks, climbing hills of extraordinary steepness with descents on the other side of equally terrifying appearance, with the constant fear of the precipice before the eyes of the driver should he miss his corner at the bottom.
>
> (Gerald Rose – A record of motor racing, London, 1909)

The crossing was full of incident both for the Gordon Bennett competitors and those in the main race. At one point a contestant named Captain Genty was negotiating his way down a very steep hill when he heard cries for help behind him. A motorcyclist's brakes had failed and he was careering down the hill towards a horrible death. As the motorcyclist passed the racing car Genty grabbed him by the collar and hauled him to safety in the car. Amazingly, eighty-five of the ninety-three cars that started that stage got through the pass, and there were no fatalities. But it was a close-run thing. Clutches, engines and brakes were tested as never before. Spectators and mechanics were called upon to push; every spare ounce of weight – mechanics, tools, fuel and in some cases even bodywork – was shed from the cars (and had to be retrieved on foot). Brakes and tyres caught fire and engines overheated. One contestant, Louis-Emile Rigolly, found he could only make the steepest hills in reverse.

Edge performed this crossing with their Napier brakes – which consisted of metal bands lined with leather, at least until the leather burned away – no longer functioning. Somehow they got the car to the foot of the pass, the vehicle suffering a terrific buffeting from the configuration of the road. However, when they got there they discovered that all their tools had fallen off the back of the body (along with the back of the body itself), leaving them without any means of effecting the tyre change they so badly needed. By now they knew victory was within their grasp, for they had seen the Panhard of their last remaining competitor, de Knyff, broken down and abandoned on the mountainside. In a feat of superhuman strength for which none of them was subsequently able to account, they managed to change the tyres without tyre levers. They limped the car to Innsbruck where, after despatching the best efforts of the French to disqualify them on a technicality, they were awarded the Gordon Bennett Trophy. The Paris–Vienna race itself was won by Marcel Renault in his Renault. The colourful Count Zborowski actually finished first on the road, but was ruled out on a timing technicality.

Leon Serpollet built this four-cylinder steam car in 1902. It developed 106bhp and propelled him to a record 75mph.

Renault's car was actually an entrant in the light car class. He had concentrated on building smaller, lighter, more nimble cars, rather than going all-out for maximum cubic capacity and brute force. It pointed the way ahead for the industry, and the lessons learned from the racing cars of 1902 found their way into the touring cars of 1903. Their other secret weapon was what Marcel called 'his special technique' for taking corners (others called it 'irresponsible behaviour' or 'driving like a maniac'). In any case he certainly went like the wind, at one stage exceeding 70mph. Thirty-three miles out from the finish he passed the then leading heavy car, Henry Farman's 70hp Panhard, to claim victory.

About this time the first tentative move was made towards the present-day concept of circuit racing when a 351-mile race was staged over six laps of a route in north-east France, south-east Belgium and Luxembourg – what became known as the Circuit des Ardennes. However, a fifty-eight-mile circuit, relatively unmarked, still takes some learning and English entrant Charles Jarrott had to follow the tyre tracks of his competitors over the initial laps. He might also have followed their dust, since this was a particular problem in this race, making overtaking a matter of blind courage, which caused many accidents. But Jarrott learned the circuit well enough to win the race.

The increased speeds of the cars made aerodynamics more important. Fernand Charron was one of the first to try streamlining his Panhard. Some of these early efforts were not entirely successful due to a failure to anticipate the consequences of doing so. The streamlined Panhards in the 1904 Gordon Bennett race suffered from overheating, because the streamlining gave inadequate airflow to the (non-honeycomb) radiator. Leon Serpollet equipped his steam-powered racing car with a body resembling an upturned boat.

Steering wheels, still a relatively new feature, even on racing cars, were increasingly raked from the vertical to improve aerodynamics. But some of the finest aerodynamics of the period – described as being forty years ahead of their time – appeared on a test track on Staten Island in 1902. Charles Baker's electric car was there for an attempt on the world land speed record (which was held until 1902 by Jenatzy's electric car *La Jamais Contente*). He was on course to take it when the crowd spilled onto the track before the run was completed and he was forced to brake fiercely from an estimated 85mph, locking the brakes and crashing, wrecking the car.

1903: PARIS–MADRID

The Paris–Madrid race of 1903 was to be the last of the great city-to-city races. The Government was wrestling with the dilemma of the growing danger of these poorly supervised, or rather, unsupervisable, races, on the one hand, and, on the other, pressure to allow them to take place from a world-leading French motor industry that employed over 25,000 workers and had earned 16 million francs in exports in 1901 alone. The race to Madrid was especially important to the industry, being seen as a way of exposing the undeveloped Spanish market to this means of transport. Public opinion was equally torn between those who were appalled at the accidents and those who relished what was the modern equivalent of a gladiatorial contest.

As we have seen, some of the cars by now had huge engines and were very fast. Average speeds of 60mph and maximum speeds of to 90mph were now commonplace (Renault was actually timed at nearly 90mph in one section of the race, in what was classed a light car). But the cars were not correspondingly manoeuvrable or quick to stop. Crowd control was also negligible on such an extended route, along which (in the case of the Paris–Bordeaux section of the race) an estimated three million spectators gathered. Wandering animals, pedestrians totally unused to vehicles moving at these kinds of speeds and slow-moving traffic on the unclosed roads made these races a nightmare. Even Monsieur Tampier, the official timekeeper of the early races who was active in their promotion, thought it appropriate to advise that, 'We do not anticipate more than four deaths, onlookers or drivers....' (*Villard*, page 89)

The entry looked disturbingly large. There were 314 initial entries of which 221 made it to the start line. All of the leading racers of the day were there, along with a number of new contenders. One that caught the journalists' eye was Madame Camille de Gamond du Gaste, described by one overwrought reporter as 'Valkyrie with a mechanical steed'. She had already taken nineteenth place on the Paris–Berlin race, and was one of the main hopes of the up-and-coming de Dietrich Company, though she was somewhat restricted by the fact that the tight corsets of the day would not let her lean forward like the male competitors to reduce the wind resistance of her body. Among others, she attracted a considerable feminist following.

As for the cars, the size and power of the engines had reached alarming proportions. The 90hp Mercedes were 12.7 litres, the 70hp Mors was 11.5 litres and the new Gobron-Brillies were rated at 110hp. On top of this, ever more ingenious ways were being found to lighten the cars, to evade the weight limits. Cars were sent for scrutineering with their bonnets removed, other bodywork stripped and tool-kits reduced to an absolute minimum. Even seats were discarded in favour of a rug on the petrol tank, (perhaps the most extreme example of this occurred in a Daimler, entered some time afterwards at Brooklands, which even dispensed with the rug, attaching the driver directly to the petrol tank with a rope).

Between 100,000 and 200,000 people went to watch the contestants leave Versailles at 3.30 in the morning. Two of the crowd were run down and seriously injured by a contestant who was still making his way to the start. The start was described by one of the entrants as 'a ghastly sight': three or four miles of straight road packed with milling crowds, cattle and dogs roaming free, the blinding flashes of photographers in the half light of dawn, railroad crossings that might or might not be open, and humpback bridges and potholes that were only seen at the last minute in the feeble headlamps of the day, if at all.

Charles Jarrott, first away in his 11.4-litre De Dietrich, describes the terrifying anarchy of the start:

> It seemed impossible that my swaying, bouncing car could miss the reckless spectators. A wedge-shaped space opened out in the crowd as I approached and...at times it seemed impossible for the car not to overtake the apex of the human triangle and deal death and destruction. I tried slowing down, but quickly realised the danger was as great at 40mph as at 80. It merely meant that the crowd waited a longer time in the road.
>
> (Charles Jarrott – Ten years of motors and motor racing, London, 1906)

The crashes began almost immediately. The American Terry hit a kerb and burst a tyre, sending him wildly out of control, crashing, already in flames, into an apple tree. Terry was dragged from the blazing car by his mechanic, as the local peasantry rushed to the scene to gather souvenirs. Crowds would gather around broken–down vehicles, creating huge hazards for the contestants following behind. Accidents and breakdowns were happening thick and fast and there were as many reports of hair's breadth escapes. The first fatality of the race came near Chatres where a young woman was knocked down and killed. Near Bonneval the Wolseley driver Leslie Porter swerved into a ploughed field to avoid a closed, and unsupervised, level crossing. He was thrown from the car, which continued into a farmhouse, bursting into flames and killing his mechanic.

Marcel Renault was the next to die, misjudging a corner during an overtaking manoeuvre, crashing spectacularly into a tree and suffering serious internal injuries. His competitor, and friend, Maurice Farman, stopped, sacrificing his own chances in the race, to administer first aid, and then went off to find further medical supplies in the nearest village. Despite his efforts, Renault died later in the night.

For once some small concession was made to road safety. In fourteen main towns the competitors were halted at its outer limits and their times recorded at an in-bound control point. They were then despatched through the town behind a bicycle, which they were not to overtake on pain of disqualification. This led them to an out-bound control point, where their time was recorded again and they could recommence the race. In smaller towns and villages anarchy ruled as usual and the local police were left to their own devices, trying as best they could to persuade residents to stay indoors as the racers passed. It did not work.

Georges Richard had a peasant in a donkey cart pull out directly in front of him. Three people were seriously injured in the collision. Louis Renault recalls running over four dogs along the route.

An appointment with death – Marcel Renault driving in the Paris–Madrid race of 1903, on the day he crashed and was killed.

Claude Lorraine Barrow was less lucky with the dog he hit; it got tangled up in his steering and caused him to crash, killing his mechanic instantly and seriously injuring himself. Barrow also died later of his injuries and, in a bizarre postscript to his death, Barrow's beneficiaries had to sue his insurance company after they refused to pay out on his life insurance. He had apparently not declared his occupation on his application form, but had simply put himself down as 'a gentleman'. The relatives argued that the insurance company should pay up since 'racing motorist' was a perfectly normal activity for a gentleman! The court agreed with them.

E.T. Stead crashed into a ditch at 70mph, ending up pinned under the car with petrol pouring over him. Madame du Gaste (who herself may have had a good chance of winning) stopped, freed him from under the car, administered first aid and waited with him for the ambulance to come. Her selflessness cost her three hours and relegated her to forty-fifth in the race.

There was even worse carnage when Tourand swerved to miss a little girl and went into the crowd. He crashed into a tree killing his mechanic, the little girl, a soldier named Dupuy who had tried to save the little girl and another spectator. Dupuy later came to be regarded as something of a martyr by those who wanted to see racing banned.

By now there were at least nine dead and an unspecified number injured. Vanderbilt put the figure higher, at fifteen dead. The carnage caused panic among both the organisers and the Government in Paris. Early in the afternoon the authorities announced that the race would be abandoned when it reached Bordeaux. The cars were not even allowed to return to Paris under their own power, but were towed to the railway station by horses. For their part, the Renault Company announced its total withdrawal from motor racing, following the death of Marcel.

The magazine *L'Illustration* described the aftermath of the race as '…a dreadful sight… machines upside down, smashed, broken in pieces: inert bodies of men killed on the spot, of the injured, moaning or unconscious…' and accompanied its account with photographs as lurid as its prose.

Meanwhile, victory in the 1902 Gordon Bennett Trophy meant that the British motor industry had finally come of age. Unfortunately its legislators had not. Under the rules of the Gordon Bennett competition, the 1903 race had to be held somewhere in the British Isles. This presented a real problem, given the country's strong anti-motoring lobby. Predictably, the British Government refused to close off any public roads for the 1903 Gordon Bennett race, so the organisers hit on the idea of running it in Ireland where the Viceroy was rather more liberal-minded. A Bill was passed in the House of Lords and a course laid out at Athy, to the background of a chorus of carefully whipped-up public support. This included the rather bizarre argument that the possibility of a potato-based fuel might hold out rich prospects for the Irish economy.

There is also a suggestion that the British team adapted the Irish national colour – green – to become British racing green, in acknowledgement of their hosts for this event. But the Napier that had given Britain its first motor racing triumph in the previous year's race had also been finished in green – what was then known as Napier green. Britain had by now gone from having no entrants, to having so many that an eliminating trial had to be held, and this was kept free of the interference of the Government by conducting it on the private estate of the Duke of Portland.

One benefit of a race being run on a circuit rather than between two cities was that it became rather more practicable (and a lot more necessary) to do something about the dust problem. This they did in Ireland by spraying the circuit with a mixture of petrol and ammonia known as Westrumite. Whilst it bound the surface and reduced the amount of dust thrown up, it also meant that what was thrown up cut the faces of tailing drivers and was so acrid it nearly blinded them. The concrete track at Brooklands was probably the first to be properly surfaced for racing, and it was not until the French Grand Prix of 1908 at Dieppe that the event could be run on a fully tar-macadamed track. Even then the standard of workmanship was such that the cars were throwing up tar-coated dust onto those behind. At a subsequent United States

One of the roads outside the Crystal Palace receives a demonstration of the dust-laying compound Westrumite in 1903.

Grand Prix they hit upon the even more radical solution for keeping the dust down: spreading eighty gallons of oil over the course.

The French also had problems with their elimination trials. The Government refused to allow a road race, but the Marquis de Dion got a Bill before the House of Deputies authorising a road race in the Argonne area. According to Scott-Moncrieff the Bill may have owed its easy passage into law to the fact that the Marquis de Dion – noted as one of the most lethal duellists of the day – let it be known that he would regard it as a personal insult if anyone raised serious opposition to it.

The German team selection for the 1903 Gordon Bennett race provides an interesting illustration of the class-based society in which motor racing used to take place. Daimler wanted to enter their new 90hp Mercedes cars with works drivers, who were in fact promoted mechanics with proven racing skills, but the rules required that they be members of the national automobile clubs represented in the competition. The Deutsch Automobile Club were appalled, insisting that their nation could not possibly be represented by 'workers', a people they considered vulgar and proletarian. They wrote, 'only ... gentleman drivers can take part in this race ... (and yet they) cannot become members of the Club...because the Club is not in a position to accept unsuitable personalities.'

A compromise of sorts was arrived at, with a new team being formed that was both technically and aristocratically qualified. However, the matter was eventually resolved in a rather more fundamental way when the Mercedes factory caught fire and all the team's cars were destroyed, just three weeks before the race. Without a works team private owners stepped in to represent Germany, the nation's hope resting primarily on Camille Jenatzy in a borrowed 60hp Mercedes road car. Jenatzy had a reputation as someone whose bravery outdid his wisdom, who drove cars that nobody else would touch, and who usually ended up broken down or in a ditch. The French, who had reliability problems of their own, banked on Jenatzy eliminating himself. But for once Jenatzy had a reliable car and was able to keep it on the track. By the time the French were alerted to this departure from his normal practice, Jenatzy was too far ahead to be overhauled.

This meant that the 1904 race would be held in Germany, which was possibly just as well since the French had banned long-distance racing after the disasters of the Paris–Madrid race. An eighty-seven-mile road circuit was identified near Frankfurt-am-Main, and the race developed into a competition between last year's winner, this time in a more powerful Mercedes, and the Frenchman Leon Thery in a Richard Brasier. The decisive factor this year was that Thery's car had shock absorbers on the front wheels making it far more stable, and Jenatzy had to settle for second place.

By 1905 the cost of running an international motor race had increased so much that the promoters had to look for the highest possible local sponsorship. For this year's Gordon Bennett race Clermont Ferrand, where Michelin tyres were made, was called upon to host the event, despite the fact that the area had nothing like enough hotel accommodation for the thousands of spectators who now attended these spectacles. The French then tried to turn the race into a French Grand Prix, but the other competing nations would have none of it.

In the 1905 French eliminating trials Panhard went all out for power, managing to develop a 120hp engine within the weight limit of 1,000kg, but they could not compete with the smaller and more technically advanced Richard Brasiers (11.3 litres, giving 100hp at 1,350rpm). Jenatzy started the race as the favourite in his new 120hp Mercedes, but he ran into mechanical trouble. A team of 16.2-litre FIATs made their first appearance, from which one of their drivers, Vincenzo Lancia, the son of a tinned soup manufacturer, would go on to become a car maker in his own right.

The administration of the race was as usual somewhat chaotic, this time because the local peasantry stole the telephone cables on which the organisers depended. Once again Thery won the race for France. This presented the organisers with a problem, because the French had already announced that they would not take part in next year's race, far less host it, for reasons explained in the following chapter. The Germans argued that speed events had served their purpose (possibly because they were no longer able to win them) and the British and American teams both decided for their own reasons not to compete for this trophy again. Thus ended the short, shambolic, but highly significant episode of the Gordon Bennett Trophy.

Of course racing continued, but increasingly on circuits and with the authorities forever trying (as they do to this day) to contain the ever-greater performance of the cars. By 1907 the authorities had dropped the weight limit in favour of a fuel consumption limit equivalent to 9.4mpg, but this did not encourage many of the competitors to reduce their engine sizes or to make any real concessions to fuel economy. Although this did result in rather more sturdy cars being entered, their extra weight meant their tyres suffered more, and thus there was a natural limit to what could be achieved in that direction. The following year the formula was changed again in favour of limits on the bore of the engine cylinders. This left it open to manufacturers to increase engine sizes by having a longer stroke, but (unlike voiturette racing) most did not do so because the high piston speeds that this required presented them with reliability problems.

What Jellinek said about the commercial imperative to take part in racing was proved correct time and again. Many manufacturers saw their sales rocket after competitive success and, on the various occasions that Mercedes tried to withdraw from the track, they saw their sales plummet alarmingly – on one occasion in 1908 to the point where they had to start laying off staff.

Leon Thery and his mechanic Muller celebrate their 1905 Gordon Bennett Trophy victory.

CHAPTER 6

THE BIRTH OF
CIRCUIT RACING

An essential feature of the racing motor car is the combination of a disproportionately powerful engine with a comparatively light and fragile carriage. The result is a locomotive which at high speed is practically unmanageable and requires almost as free and uninterrupted a course as an express train…The slightest obstacle in its way is equivalent to imminent risk of death to the occupants of the car as well as to any unfortunate persons who may happen to cross its path.

(The Times, 26 May 1903)

Poor old England – the cradle of sport as she used to be called and now she is nowhere. The time is surely come when England should no longer lag behind the rest of the world but take her place at the very forefront, if possible, and reassert herself as the arbiter of sport.

(Hugh Locke-King laments the lack of facilities
for motor racing in England)

Whilst city-to-city racing largely disappeared in mainland Europe, new circuits continued to emerge – among them the world's first purpose-built racing circuit. Road racing did continue, however, if to a lesser extent. A wealthy Sicilian winemaker, Caviliere Florio, set up a ninety-three-mile course through the mountains of his native island in 1906. Entrants were discomforted to find that during practice bands of local bandits who had never been unduly troubled by the rule of law were taking pot-shots at the contestants with rifles throughout the course. When complaints of this reached Florio he adopted a truly Sicilian solution: he bought in the leaders of the island's three main gangs of brigands, appointed them as stewards and sent them out, heavily armed, with instructions to enforce the 'no shooting the drivers' rule on the lesser bandits by whatever means they felt necessary. The race went off without incident (at least, as far as the drivers were concerned) and the Targa Florio was born.

One significant – but at that time generally unknown – visitor to the event was an English landowner, son of the Seventh Baron of Ockham, Hugh Locke-King. He was shortly to become one of the most important – if not best known – names in British motor sport. His claim to fame will emerge shortly.

In the same year, having superseded the Gordon Bennett races, the French Grand Prix was held by most people to have been the first proper Grand Prix, though the Automobile Club de France maintains that the first such event was the Paris–Bordeaux Race of 1895. The problem with the Gordon Bennett Trophy, from the French point of view, was that it gave them, with their multitude of competitive manufacturers, only three entries, the same as their most negligible competitor. As the French presented it, their revised rules would give interested manufacturers throughout the world – most of whom happened to be French – an equal chance of winning. Their proposal was that the first fifteen French cars in the Gordon Bennett trials should have a place in the Grand Prix, along with six each from Germany and England, and three each for Austria, Italy, Switzerland, America and Belgium. The proposal was first floated in 1905. Initially, it was dismissed by foreign manufacturers and by those concerned about the number of cars racing at one time. The demise of the Gordon Bennett series left the field open for someone to organise a new series of elite competitions, and the Automobile Club de France seized it. Under their new rules there would be no control points and no outside help for the competitors.

Once again the motor clubs were dependent upon local authorities to subsidise them. In this instance the authorities in an area of Sarth, centred on the town of Le Mans, 130 miles west of Paris, came up trumps (perhaps it is a coincidence that the Bollee automobile factory was based there). The starting straight was a section of the Route 23 Paris road, and the circuit comprised three legs of about twenty miles each. Forty miles of barricades were erected to withhold spectators, along with a grandstand and a tunnel which allowed spectators to venture inside the circuit. Two sections of the circuit, however, still consisted of wooden decking.

BROOKLANDS AND CIRCUIT RACING

> The face of the landscape has changed completely, since ground was first broken last October – and in the opinion of the majority of Weybridge residents, for the worse. The fact remains, however, that there is in existence a track for motor racing that from a theoretical point of view is as nearly perfect as human ingenuity can conceive…those who saw the site of the track last October and again yesterday could not but be struck by the enormous amount of work that has been accomplished.
>
> *(The Times, 18 June 1907)*

Up until 1907 there was no such thing as a dedicated motor racing track. It was the wealthy property owner, Hugh Locke-King, who was persuaded to invest £150,000 of his own money (well over £10 million in today's terms) in constructing a track in the middle of the Surrey countryside, setting aside 300 acres of his own estate at Brooklands. While watching a motor race in Brescia, Italy, he asked why the international list of entrants contained no Britons. One of the reasons he was given was that they had nowhere to practice.

Brooklands was no ordinary track: 100ft wide and 2.75 miles long, with a half-mile straight, it had to divert the River Wey in two places and involved the clearance of some thirty acres of woodland and two farms. It was also built on the site of a 2,500-year-old Iron Age village and a Roman settlement. But the most striking feature of the track was its banked corners, so steep at the top that it was almost impossible for a pedestrian to climb up them. Cars going too far up the banking at too low a speed ran the risk of sliding back down (possibly onto the cars they were overtaking). Designed by Colonel H.C.L. Holden of the Royal Engineers, the banking meant that, from quite early on, laps in excess of 100mph were possible and, furthermore, even at such speeds, steering would not need to be applied on the banked corners. It was in fact designed to allow for cornering speeds of what was then an unimaginable 120mph. But the very earliest record to be set there was by the redoubtable S.F. Edge, who single-handedly drove his 60hp 7.6-litre Napier around the track for twenty-four hours on 29 June 1907, covering 1,581 miles at an average of 66mph – more than a mile a minute.

As *The Times* (quoted above) intimated, Locke-King also had to contend with the rooted opposition of his neighbours, as the diary of a local resident, Lady Mary Monkswell, states that:

> …they have been building this awful motor track and are so hated by their neighbours, many of whose houses they have simply ruined, that hardly anyone will speak to them…
>
> …The neighbours thirst for Locke-King's blood.

> (A Victorian Diarist – Mary, Lady Monkswell – edited E.C.F. Collier, 1946)

This local opposition secured some concessions such as noise reductions and a ban on night racing, but they could not stop the circuit being built. Up to 2,000 labourers were employed in its construction, which took just nine months. In the course of its construction, they felled thirty acres of woodland, moved 350,000 cubic yards of earth and laid 200,000 tons of cement, most of it mixed by hand.

An opponent of the scheme or not, Lady Monkswell allowed a friend of hers to drive her round the circuit, and recalled: 'She put it to 43mph – I felt my eyes pressed in by the air at that terrific speed, and I could hardly breathe.' As they rounded the banked corner she also experienced the thunderous sensation of an 80hp Darracq passing more or less directly over their heads at 90mph.

The circuit opened on 17 June 1907, the world's first purpose-built track. There were prizes totalling £4,450 for the drivers in the first meeting and admission charges ranged from 2s to the £1 charged to the elite for admission to the reserved lawn. A stately procession of the great

The opening day parade at Brooklands in 1907 that soon turned into an out-and-out race.

and the good in their cars, led by Mrs Locke-King, was supposed to open the proceedings. S.F. Edge describes what happened next:

> Mr and Mrs Locke-King led the way in their 70hp Itala …There was no attempt to indulge in speed work until Mrs Locke-King let her car out a little.
>
> Lord Lonsdale gave chase and within a few minutes we were all tearing around the track as hard as we could go. There were about sixty of us, all rushing along for dear life.
>
> I do not know what speed we were doing, but suddenly there appeared Warwick Wright in his racing Darracq. He passed us all as though we were standing still and we heard afterwards that he was touching 85mph…
>
> (Johnson, page 13)

Brooklands was not an overnight success. Locke-King had built grandstands for 30,000 people and accommodation for a total of 250,000 spectators, but attendance at the opening event was disappointing at around 4,000. It would soon get worse. The second meeting clashed with the nearby Sandown Park horse race. Despite reducing the minimum admission price from 2s 6d to 1s, only 3,000 people turned up. Some reported finding the racing, taking place on large expanses of concrete rather than the suicidally intimate public roads, boring. Certainly that was the view of the local paper after the first meeting:

> On Saturday last residents of Weybridge and districts adjacent to the Brooklands motor track were favoured with a foretaste of what they may expect for the next few years, or, at any rate, till the public curiosity as regards the new sport dies the natural death which probably quickly awaits it. What, after all, can there be in watching the endless circling of highly developed machines round and round a circumscribed area?
>
> (*Surrey Herald*, June 1907)

But one place where the new venue was welcomed was *The Autocar* magazine, where the opening event was the subject of an eight-page picture-spread and five pages of (largely supportive) readers' letters. Despite a shaky start, Brooklands gradually became the place to see and be seen for the next thirty-two years – 'the right crowd and no crowding' was the slogan used to describe its ambiance – and William Boddy, the circuit's biographer, described it as 'a gentleman's estate where motor cars just happened to race'.

The course featured a number of innovations: firstly, there were six linked telephone boxes around the course enabling marshals to stop the race at any time in an emergency; and secondly, an electronic timing device was installed on the track, supposedly accurate to within 1/1,000 second. This latter equipment launched a controversy in 1908 that continued for decades, after recording a lap speed (as distinct from a land speed record attempt) of 121.64mph for Felice Navarro's FIAT, a time some 20mph faster than anything hitherto seen at the circuit and not exceeded by any other racing car until well after the First World War.

Nobody knew how to run a circuit race meeting since there had never been anything quite like it before, so they modelled it closely on horse-racing events. Drivers wore smocks in their owners' colours (numbers on the cars were considered vulgar), cars were handicapped by weight and price and the races were all called 'plates' in the horse-racing manner. In line with the sporting elitism of the day, drivers were not allowed to use the same bar as club members. Also in keeping with the horse-racing theme, they employed a former Jockey Club starter, Hugh Owen, to begin the races. There were even complaints by Charles Jarrott about the automotive equivalent of racehorse doping – that competitors were feeding pure oxygen into their engines.

The racing season also kept initially to the same start and finish date as the flat racing season. This winter hiatus resulted in something of a cash-flow problem for the small army of mechanics, many of whom set up in business there to cater for the needs of their rich but nonmechanical customers, and they were forced to service buses during the winter to make ends meet. On non-race days the track was open to all-comers for a fee of 10s. Individual car owners took advantage of this, as did many manufacturers, who could subject their cars to testing at speeds unattainable elsewhere.

In an attempt to appease the neighbours Brooklands introduced minutely detailed specifications for exhausts and silencers. Despite this residents took Locke-King to the High Court, complaining of the noise from the cars and the nuisance of the crowds. It cost Locke-King £7,000 in legal costs and the price of a new access road.

Brooklands soon became as famous for pioneering aviation as for motor racing. Flying schools were established, many of the names that were to be famous in the history of the Royal Flying Corps (AVRO, Vickers, Sopwith and Bleriot) setting up manufacturing facilities there. In addition to this, from 1911 there was an office from where flights to various destinations could be booked, making it a very early version of an airport. The cars still had the edge for speed over these early aircraft: in 1911, a Hanriot monoplane and a Singer racing car raced each other round the circuit, entering the finishing straight side by side (or top to bottom!). The aircraft then laboured down the straight at a flat-out 40mph while the Singer sailed past at 78mph.

So fast was the circuit that some of the early world land speed records were set there. In 1909 Victor Hemery broke the 200kph barrier there (recording 202.65kph, 125.92mph) in his 21.5-litre 200hp Blitzen (Lightning) Benz. This car (designed by Hans Nibel, later to produce the legendary Grand Prix racing Mercedes cars of the 1930s) was ultimately capable of 140mph with its engine turning over at just 1,400rpm.

The ultimate lap record was later set in 1935 at 143mph by John Cobb in his 24-litre Napier Lion aeroplane-engined car. At the other end of the spectrum, a little Singer Ten of 1912, a budget car of 1,096cc costing just £195, managed to lap Brooklands at 64mph, while a 1913 Model T Ford recorded a lap speed of 56.75mph.

The spectacle of high speeds, however, was clearly not considered to be sufficient on its own to attract large crowds. Among the other pre-war diversions, promoters offered their audiences 'blindfold' racing, relay races, ladies' races, obstacle races – with dummy pedestrians for the competitors to weave among (or possibly over) – 'veteran' car races (defined in 1911 as pre-1908 cars, or cars over three years old), FIAT taxi cab races (with a winning average speed of 32.96mph) and tyre-changing races. In these latter, competitors were required to change the offside rear tyre. The use of detachable wheels and rims was barred, so it was literally a case of replacing one tyre with a spare, which then had to be inflated by hand. How much of a spectacle this made for the onlookers is open to question, but tyre wear generally was a major element in the early years, resulting in many accidents and retirements. The Blitzen Benz, referred to earlier, could not manage more than fifty-eight miles on a set of tyres, and on one occasion when one blew at high speed it initiated a skid measured at 1km. Tyre changing encouraged the development of slick pit techniques that anticipated modern Formula One technology. The quick-lift jack had made its appearance by 1911 in the form of a plank pivoting on a log. During a record-breaking attempt that year the Sunbeam pit crew got the time required for a tyre change, refuelling, and the adding of water, oil and grease down to eighty-seven seconds.

The track soon became a magnet for the motor racing fraternity worldwide. But then the war came, and other aspects of the circuit took prime importance over motor racing: the Royal Flying Corps commandeered Brooklands, and their heavy vehicles lumbering up and down the circuit meant that it was never quite the same again after the war. But one further record was to be broken before the end of the war: the Vickers Vimy bomber which Alcock and Brown used in their pioneering flight across the Atlantic was built at Brooklands.

One of the consequences of circuit racing replacing road racing was that the need for a mechanic to travel with the car ended. Proper accommodation for the mechanics was first provided on the circuit at Dieppe for the 1908 French Grand Prix. However, they were built below ground level and the name coined for them was one that stuck – the *pits*.

Two cars compete in the wide open spaces of the Brooklands track.

The 1908 Dieppe Grand Prix, showing the semi-underground 'pits'. Garcet in his Motobloc comes in for repairs.

At the lower end of the sporting ladder club events were now starting to be organised in Britain for the racing motorist. There were hill climbs, such as the Sunrising Hill Climb, organised by the Midland Automobile Club, and sprint trials at Blackpool and Bexhill. These generally took place on public roads, which were subject to the national speed limits. Sympathetic police were asked to turn a blind eye, the organisers 'requesting' other traffic give a wide berth to the competitors. To avoid inviting prosecution for doing several times the legal speed limit, the clubs tended to avoid publishing the winners' times or speeds, simply listing the number of seconds the runners-up were behind the winners. The Isle of Man, used for the British elimination rounds of the 1904 and 1905 Gordon Bennett races, became the home for a series of races, not for stripped-down racing cars but for genuine four-seater touring cars – the Tourist Trophy.

A week after the starting gun for the First World War was fired at Archduke Franz Ferdinand in Serajevo, the starting flag was dropped on the 1914 French Grand Prix. This was a race that has been called the greatest Grand Prix of all time, and one which marked the end of an era in European motor racing history. The twenty-three-mile circuit near Lyons was described as near perfect, and the huge crowds invested the event with all the glamour that *Belle Epoque* France could provide. Thirteen teams representing every motor car manufacturing country except America entered forty-one cars. The race was run under new capacity (4.5 litres) and weight (1,100kg) limits. The French Delage and Peugeot teams, who had recently won a clean sweep at Indianapolis, were favourites, and in Georges Boillot Peugeot had what was generally regarded as the world's greatest driver. But the French feared the threat posed by Mercedes, while the British (Sunbeam and Vauxhall) and Italian (FIAT) teams were not to be ruled out.

In fact, the Mercedes team saw the race as a prelude to war by other means, and made unprecedented preparations for it, arriving at the course for practice three months before the actual race. Many aspects of their cars' engines drew upon their research into aeroplane engine manufacture. But, unlike many of the entrants, the Mercedes still relied on a single camshaft and had brakes on the rear wheels only. Also, one of their leading drivers, Christian Lautenschlager, the winner of the 1908 Grand Prix, was derided as being too old to compete.

From the start Max Sailer's Mercedes set off at a blistering pace, designed to wear down the French rivals. Unfortunately, Sailer also wore himself down, retiring on the fifth lap with a mechanical failure. Boillot emerged in front, as his rivals fell around him, driving (it was said) at 'fabulous, fantastic speed'. Meanwhile Lautenschlager quietly worked his way through the rest of the field until, after six hours of racing, he was only fourteen seconds behind Boillot. A lap or two later Lautenschlager took the lead. Then, on the very last lap, Boillot's Peugeot blew a valve and coasted to a halt. Mercedes took the first three places, and the last of the great pre-war motor races came to an end. The French band could not bring itself to play the German national anthem and, in the newspapers, coverage of the event took pride of place over the world war that was unfolding before them.

LAND SPEED RECORDS

You have every sensation of being hurled through space. The machine is throbbing under you with its cylinders beating a drummer's tattoo, and the air tears past you in a gale. In its maddening dash through the swirling dust, the machine takes on the attributes of a sentient thing…no man can drive faster and live!

(Barney Oldfield describes setting a land speed record of 131mph

in a 21.5-litre Benz in 1910)

The world land speed record had its origins in an 1898 wager between Count Gaston de Chasseloup-Laubat and the Belgian, Camille Jenatzy, known for his red hair and beard as the Red Devil. Chasseloup opened the bidding by covering a kilometre in fifty-seven seconds (an

Gaste in a Gregoire and Vallee in a Werner compete at the hairpin at Dieppe in the 1908 Grand Prix des voiturettes.

Christian Lautenschlager on the straight in his Mercedes in the Grand Prix at Dieppe in 1908. He was later the hero of the 1914 French Grand Prix.

average speed of 39.244mph) in his electric car *Jeantaud*. Jenatzy responded the following year in another electric car, this time built of an early aluminium alloy, partinium. *La Jamais Contente* achieved an average speed of no less than 62.167mph, more than 100kph. As an aside, Jenatzy went on to suffer an untimely and bizarre death behind the steering wheel, but one entirely unconnected with his love of speed. He was one of a group out hunting wild boar in the Ardennes in 1913, and decided, for some inexplicable reason, that it would be highly amusing to hide in the undergrowth and make wild boar noises. His fellow hunters duly shot him and, on discovering their mistake, carried him back to the cockpit of his Mercedes to die.

Electric and steam cars dominated the early years of the land speed record. It would be 1902 before a petrol-engined vehicle (a 60hp 9.2-litre Mors belonging to the American millionaire William K. Vanderbilt and driven by the Frenchman Henri Fournier) took the record, raising it to 76.583mph. Land speed records were still being set in unexpected places. Rigolly became the first man to exceed 100mph on the Promenade at Ostend and, just before the First World War, L.G. 'Cupid' Hornstead recorded a two-way record of 124.1mph on the Brooklands circuit, at the wheel of a 21.5-litre Benz.

Even so, steam came back for one last hurrah in 1906 when an American called Fred Marriot recorded 121.544mph in a streamlined Stanley steamer. This had a 3.1-litre two-cylinder steam engine (equivalent to a 12.25-litre internal combustion motor) that developed 120hp at a boiler pressure of no less than 1000lb/sq.in. The car actually exceeded 200kph (124.27mph) for part of the attempt, but was not able to sustain it for long enough to put this landmark into the record books. Marriot's 121mph record for a steam car did, however, become the longest-standing land speed record, surviving well into the twenty-first century.

Camille Jenatzy, the red-haired racing driver, world land speed record breaker and (fatally for him) wild boar impersonator.

Sunbeam were one of many cycle manufacturers drawn into car production. From 1909 the talented French designer Louis Coatalen was their guiding light. Their output included racing cars like this 1914 example, and aeroplane engines, but they won particular fame with their land speed record breakers.

ENDURANCE

The new outlets for the adventurous motorist, after the demise of the inter-city road races, were the long-distance endurance trials. One of the first and most ambitious of these was undertaken, solo, as early as 1902, by a naturalised Briton called Dr Lehwess. He commissioned a specially built yellow Panhard which he named *Passe-Partout* after the character in Jules Verne's *Around the World in 80 Days*. It cost him the enormous sum for the day of £3,000, and he set off in it to emulate Passe Partout's circumnavigation of the globe. He left London with a flourish in April, but was forced to abandon the attempt with rather less ceremony at somewhere called Ninji-Novgorod in Russia the following September. His car was so thoroughly bogged down by then that he had to abandon it for the winter. The following year the London motor dealer Charles Friswell travelled to Russia and paid the locals to dig the vehicle out from under 3ft of snow and an unspecified quantity of mud. It was returned ingloriously to London by train.

The first competitive international endurance trial took place in 1907, with the French newspaper *Le Matin* sponsoring a long-distance race from Peking to Paris. It was hardly a race in the mould of Paris–Madrid; for parts of the course the cars had to be ferried by teams of coolies or mules over trackless mountains, rivers and other obstacles impassable by the cars under their own power. There was a feeling that there had been a surfeit of circuit racing, which was influencing car development in the direction of greater power and speed. But this was a new and wider type of challenge, designed to prove that 'as long as man had a machine he could do anything and go anywhere'. For their part, the Chinese authorities would rather have seen this influx of Westerners go anywhere else, fearful as they were that 'they would cause an

Motoring is introduced to some new and exotic outposts during the 1907 Peking–Paris race.

upheaval in the popular mind' and 'spread everywhere the fatal germs of Western corruption'. As an aside from this, motoring seems to have been generally slow to catch on in the Far East. As late as 1908 there were only twenty cars to be seen on the streets of Tokyo. Nevertheless, the contest was won by Prince Scipione Borghese in a 7.4-litre Itala, who covered the 10,000 miles in sixty days.

Six months later the *New York Times* and *Le Matin* jointly sponsored a race from New York to Paris, taking the scenic route through San Francisco, Alaska and Asia. Six cars headed westward from New York. They were equipped for every eventuality. One of the French entrants, a de Dion, had steel-studded wheels for use on ice, skis to replace the front wheels on snow, flanged wheels for running on railway tracks where there were no roads and a mast and sails for where there was no petrol. The catalogue of problems they encountered en route was endless and sometimes unexpected. One driver was fined for frightening a horse, while another crew came under suspicion of spying. They were buried in snow, overwhelmed by floods and covered in blisteringly hot sand in Death Valley. The section through Alaska had to be abandoned, being deemed impassable even by the promoters' ambitious standards. In the event, it was perhaps surprising that four entrants made it as far as Siberia, and two – a German Protos and an American Thomas Flyer – got to Paris. The American car's victory – which took 168 days – made a worldwide celebrity of its driver, George Schuster, and boosted Thomas' sales by 27 per cent.

But, fearful as conditions were in China and Siberia, the real test was crossing the United States. A Mr and Mrs John D. Davis were two of the early pioneers to attempt this, and they got as far as Detroit. No trace of them was ever found. Thirty-six other unsuccessful attempts at a crossing America had been made before the first success in 1903. The appropriately named Horatio Nelson Jackson completed this journey in sixty-four days, longer than it would take for the winning entrant to go from Peking to Paris.

But Horatio was a positive speed demon compared to a German lieutenant called Graetz, who, in 1909, took no less than nine months to drive across Africa. One major reason for his delay was that, whenever he came to a river, of which there were many, he had to recruit the local people to build a raft for him and his car.

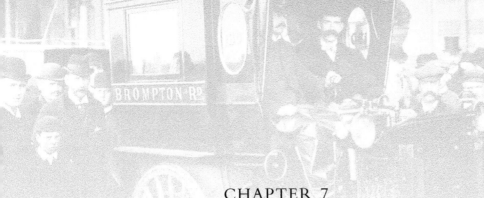

CHAPTER 7

MOTORING PIONEERS

This would be a very long book indeed if it were to try and give an account of all those who contributed to the early development of the motor car. Those seeking such coverage should turn, for example, to Georgano's authoritative and comprehensive work (listed in the bibliography). What follows is a short and selective account of some of the pioneers, and the contribution they made.

HERBERT AUSTIN (1866-1941)

Born in Buckinghamshire, Herbert Austin emigrated to Australia at the age of eighteen. There he met Frederick Wolseley, the proprietor of a company making sheep-shearing equipment, which Austin joined as an engineer, having taken night classes in Melbourne. In 1889 Austin and Wolseley moved back to Britain, where they opened a factory in Birmingham. They soon added bicycle components to their range of products. Austin saw horseless carriages at the Paris Salon in 1895 and immediately began building one of his own. It was modelled on the Bollee voiturette, except for a most unusual horizontally opposed two-cylinder engine of his own design in which the combustion took place in a separate chamber to that containing the pistons. It did not go into production, however, for two reasons: first, the parts of it that were good were not original (in that it probably infringed a lot of Bollee's patents); second, the part of it that was original (the engine) was not good.

He enjoyed greater success with his second model, a three-wheeler with a single-steered front wheel. It was well received at the National Cycle Show in December 1896 and later successfully completed a long testing run from Birmingham to Rhyl and back. One of the Austin-designed Wolseley cars also made a creditable showing in the 1900 1,000-mile trial, and orders began coming in. The engineering firm Vickers bought out Wolseley's car-making interests in February 1901 and provided the capital needed to go into full-scale manufacturing. This was no doubt influenced by Austin's friendship with one of Vickers' directors, the motoring enthusiast Hiram Maxim (of machine gun fame). But relationships with Vickers soured, and Austin was eventually forced to resign, ostensibly over his stubborn support of the horizontal engine at a time when conventional wisdom favoured the vertical variety. However,

A 1910 Austin.

there may have been other forces at play, because in fact Vickers acquired another company prior to Austen leaving, that of John Siddeley, with the aim of producing a vertical-engined car. Also, when Austin set up to make cars in his own name, his first model had a vertical engine. He had acquired a former printing works at a place called Longbridge, just outside Birmingham, and the Austin Motor Company was launched in December 1905. In 1913, just a year before the outbreak of war, the German firm Krupp are thought to have financed an expansion of his company.

Austin's first solo creation was a four-cylinder 5,182cc model based on the then-current French-designed Gladiator and Clement models (Austin's strengths lay in good engineering and styling, rather than original design). Austin sold thirty-one cars in that first year, rising to about 200 by 1910. In the meantime he managed to enter two 9,657cc 100hp cars into the 1908 Grand Prix. They finished a modest eighteenth and nineteenth, but at least they were the only British entrants to finish. In contrast to these monsters, from 1910 Austin also built a single-cylinder 7hp model (actually a Swift with an Austin radiator). By 1912 his annual output was up to 1,100 cars. The company went public in 1914 with a nominal share capital of £650,000.

Wolseley, meanwhile, continued to flourish. They had sold 327 cars in the year Vickers took over, and this rose to 800 by 1903. This made them Britain's largest car maker, a position they held until the First World War, though they were less profitable than Austin.

AMEDEE BOLLEE – PERE ET FIL (1844-1917) AND (1867-1926)

The most important steam carriage builder of the 1870s was a French bell-founder, based in a place whose name resonates like one of his bells throughout the history of the motor car – Le

Mans. Amedee Bollee saw some steam carriages at the Paris Exposition of 1867, and decided to build one of his own. Working on it in his spare time at his family's workshops, his first vehicle was completed in 1873. Called *L'Obeissante* – The Obedient One – it looked, as Burgess-Wise put it, like 'a perambulating bandstand', but included many advanced features such as independent front suspension, a geometrically correct steering system (or at least what would have been geometrically correct, had the chains that operated it not been prone to stretching) and separate V2 steam engines powering each of the back wheels, doing away with the need for a differential. It could cruise at 19mph and had a top speed of 25mph.

In 1875 Bollee decided to take it on the 135-mile journey from Le Mans to Paris. The bureaucracy involved in this journey was somewhat at odds with the rather more laissez-faire attitude later French governments were to adopt in relation to the motor car. He had to obtain specific government permission for the journey and give three days' notice of his intended route. At the boundary of each department's domain through which he passed he would be met by an inspector of bridges, who would quiz him on the likelihood of his machine doing harm to their structures. Despite preparations, Bollee claimed that he collected a total of seventy-five separate traffic penalties on his journey. But all was forgiven when he entered Paris in triumph. The car was a sensation.

Bollee followed it up with a new model called *La Mancelle* – The Girl from Le Mans – which was much more recognisably car-like, with a front-mounted three-cylinder steam engine contained within a bonnet and a shaft drive to the differential, and then a chain-drive to the rear wheels. It was exhibited at the 1878 Paris World Exhibition and attracted great interest. Demonstration runs were given in Paris and replicas were offered for sale at a starting price of 12,000 Francs. Some fifty examples were manufactured. Among his other models was a commercial vehicle, the *Marie-Anne*, developing 100bhp and able to tow a thirty-five-ton load up a 6 per cent slope, and the nearest thing, in its day, to a sports car, *La Rapide*, capable of

An early Bollee steam carriage, *La Nature*, dating from 1883.

Harrods were early converts to delivery vans with engines. This one took part in the 1896 Emancipation Run to Brighton.

36mph. But by 1880 Bollee had virtually quit the commercial steam carriage business, and the few that were made after this date were largely for family use.

It was his sons, Amedee (junior) and Leon, who made the transition to internal combustion. Amedee (junior) produced a lightweight two-seater steamer in 1885 that was capable of 25mph, and which attracted the attention of Armand Peugeot. But Amedee's interests lay more in the internal combustion engine and, by 1896, he had a two-seater car in production, one of which the brothers drove in the Paris–Marseilles race of that year. A licence to manufacture these was sold to De Dietrich Company the following year – a practice that was to follow many of the products the brothers developed.

It was the Bollee voiturettes that were the first to finish in the 1896 London to Brighton run, and for which Harry Lawson bought the English manufacturing rights. Also, as we saw, Herbert Austin based the design of the first 1895 Wolseley car closely on the Bollee. Leon then produced a four-wheeled light car and sold the manufacturing rights to Alexandre Darracq for £10,000. This proved relatively popular until it was overshadowed by the products of Louis Renault.

Next, in about 1903, came a four-cylinder car whose main claim to fame was its extremely silent running – in marked contrast to his three-wheelers. A group of English financiers paid £42,000 for the English manufacturing rights to this model, but for some reason did not take them up. Between 1906 and 1911 the Bollee 45/50hp touring car sold in England, in chassis form, for £50 more than the equivalent Rolls-Royce Silver Ghost. Leon Bollee's other claim to fame was that, in 1908, he collaborated with Wilbur Wright in the development of engines for his flying machines.

Morgan built his first three-wheeler in 1908/9 and derivatives of it continued in production until 1952. Some were capable of over 80mph and they soon made their mark in racing, setting a one-hour cycle-car record of 58.96mph at Brooklands in 1912. The model shown dates from 1926 and has a 1,206cc engine.

ALEXANDRE DARRACQ (1855-1931)

Alexandre Darracq made his money from the Gladiator Bicycle Company, which he sold and used to finance the setting up of two car builders. One, in partnership with Adolphe Clement, traded under the names of both Gladiator and Clement. These briefly got involved with Harry Lawson in 1896, but by 1902 Gladiators were producing over a thousand cars a year, four-fifths of them sold in England through S.F. Edge's company, who also marketed Napiers. The Gladiator Bicycle Company was taken over in 1909 and the brand name had disappeared by 1920.

The company that kept Darracq's name is perhaps of greater interest. Not only did the marque become famous in its own right, in its native France and abroad, it also helped set up overseas companies that were themselves to become household names. In 1902 they arranged for a German, Adam Opel, to manufacture their vehicles in his home country under the name Opel Darracq. In 1905, they set up a British manufacturing base under their own name and, in Italy they began manufacturing in Milan under the name Societa Italiana Automobili Darracq in 1906. This latter company was never successful and closed in 1909, but the company that took it over had a rather longer track record. They were the Anonima Lombarda Fabbrica Automobili, or ALFA for short, soon to become Alfa Romeo.

Darracqs made their name in racing. On at least two occasions in 1904/05 they held the world land speed record, which by then had been pushed just above the 100mph mark. The second occasion was with an early V8. In 1906 they won the Speed King title at Daytona Beach, reaching 122.45mph, and they won the Vanderbilt Cup in 1905 and 1906. Among the

An Opel track racer dating from 1902. Opel also bought the manufacturing rights to Lutzman, a close cousin of the early Benz cars.

notable racers who drove Darracqs were Louis Chevrolet, Victor Hemery and Vincenzo Floria, the founder of the Targa Floria. Darracq sold out to British owners in 1913, and in 1935 (by now part of a Sunbeam Talbot Darracq conglomerate) became part of the Rootes Group.

COUNT ALBERT DE DION (1856-1946) AND DE DION BOUTON

Albert de Dion was born in the family chateau near Nantes, part of a noble family that could trace its origins back to a thirteenth-century crusader. In his youth he was regarded as something of a playboy and a favourite among gossip columnists, not to mention a noted duellist. He was also described as having 'an ignobly eccentric part-time obsession with steam engines', and nearly brought ruin on the family with a failed attempt to sell his idea for an improved version of the then-current design to the French Navy. His father, the Marquis, even went so far as to obtain a court order preventing him from having access to the remainder of the family fortunes for his irresponsible schemes.

It was another steam engine, this time a glass toy in a gift shop window, which led to the historic De Dion Bouton partnership. De Dion was so taken with this toy that he tracked down the makers. They turned out to be Georges Bouton and his brother-in-law, a man named Trepardoux, and De Dion persuaded both of them to come and work for him, developing a horseless carriage among other things. By the end of 1883 they had developed a marine boiler for producing superheated steam and their first four-wheeled two-cylinder steam carriage. Theirs were among the first motor vehicles to use the improved pneumatic tyres developed by Michelin in 1892.

It was with a steam carriage, consisting of a tractor unit pulling the rear half of a phaeton, which De Dion entered in the 1894 Paris–Rouen race (and, as we saw, won on the road,

De Dion Bouton, founded in 1883 as builders of steam vehicles, were possibly even more famous as engine designers than they were as car manufacturers. By 1902 over 30,000 of their petrol engines were in use by 140 makers throughout the world.

notwithstanding the arcane rules that decided otherwise). This tractor unit went on to have a number of other applications, among them powering commercial vehicles and buses capable of carrying up to thirty-five people. But by the late 1880s De Dion was already starting to turn away from steam as the source of power, moving instead towards petrol. De Dion took out his first patents for 'detonating motors cooled by water' in 1889. He managed to win over an initially reluctant Bouton, but Trepardoux was an implacable opponent to internal combustion (and did not like De Dion much either), and he resigned the partnership in January 1894. De Dion developed a number of revolutionary internal combustion engines, including a twelve-cylinder radial engine, which would have been ideal for aircraft, had they been invented at the time.

But the company's greatest gift to the development of the motor car was the high-speed engine. The prototype was only 137cc and ran at the incredible rate for the day of 3,000rpm. The engine size was increased by stages to 402cc, and the production models slowed down to 1,800rpm, which was still a huge advance on what had preceded them. It was very light, due to the use of an aluminium alloy in its construction, but its real secret lay in a system of electric ignition, driven off a cam, that provided fast and accurate opening and closing of the main electrical circuit. Further development of the engine led in 1895 to a yet larger model, which they fitted to a tricycle. Various forms of this engine were produced, developing between 0.5 and 8hp with a power to weight ratio of around 25lb/hp. No less than 21,800 tricycles using this engine had been sold by June 1899.

The first De Dion Bouton four-wheeled car appeared in 1898. It was essentially an enlarged tricycle, with the passenger sitting between the two front wheels. A voiturette followed in 1899. Their cars sold in huge numbers – something like 20,000 single-cylinder cars were made, and many still survive. Their engines were even more successful, being used in around 140 different makes of car around the world. By 1911 the company had sold more than an estimated 150,000 engines, most of them for fitting into other makes of cars. Some of these, like the prototype Bianchi 'car' of 1898, were no more than two bicycles joined together, with a De Dion Bouton engine strapped between them. Among the more successful models powered by these engines was Louis Renault's first car. The De Dion motor tricycles also enjoyed some success in competition, winning their class in the 1896 Paris–Marseilles–Paris trial. Their power output of around 7bhp per litre was not improved upon significantly until 1908.

However, a single-cylinder four-stroke engine like the De Dion was never going to be as smooth-running as a multi-cylinder model. The compression stroke, on which there was the greatest resistance, was the one furthest removed from the power stroke, so the car tended to proceed in a series of jerks. In 1905 John A. Prestwich of Tottenham (better known by his initials – JAP) developed a V-twin engine in which the power strokes of the two cylinders were staggered, giving a markedly smoother run.

Over and above his manufacturing activities, De Dion was a great champion of the cause of motoring. He helped organise the 1895 Paris–Bordeaux–Paris race and founded the Automobile Club de France that ran many of the races in the years that followed. He arranged for the publishing of an improved set of road maps for French motorists, but, after complaints that this would give too much publicity to his products, he handed them over to another company, which is why the French today have *Cartes Routiere et Touristique Michelin*. He succeeded to the title of Marquis in 1901, and shortly thereafter entered politics where he enjoyed a popular career until his retirement in 1941. Not surprisingly, one of his great causes was improving the state of the nation's roads.

Unfortunately, the De Dion Bouton Company did not do so well. They were a huge concern by 1912, employing 4,000 people, but lost their way after the First World War. The two principals in the company could not maintain their earlier pace of change as they got older, and there were no successors to give the firm new impetus. They lost their mass market to Renault and Citroen, and their luxury market to the likes of Hispano Suiza, and car production ceased in 1932.

FREDERICK WILLIAM LANCHESTER (1868-1946)

The name of Lanchester is spoken with awe in veteran car circles. Laurence Pomeroy, who made no small contribution to the development of the car himself, said that Lanchester was 'a colossus, a man of imposing stature and of giant intellectual capacity'. Someone else, less enamoured of the car, described him as 'one of the most accomplished gentlemen ever to be wasted on the motor industry'. Like Daimler and Benz, he served his apprenticeship with a gas engine manufacturer, in his case in Birmingham. Also, like many of his contemporary pioneers, he saw motor cars on display at the Paris Exhibition of 1889, and by 1893 he had built a high-speed (for the day – 800rpm) petrol engine which delivered 2hp. In 1895 he began building a car which emerged in prototype form in 1897 as one of the most original vehicles of its day.

It owed little to any of the vehicles then in existence, each part being designed from first principles, using the best science of the day. Some of his innovations are detailed in various parts of the book – his improved tiller steering, high-tension ignition, an ingeniously balanced

Frederick Lanchester in his 1897 car.

and hence smooth-running engine, air-cooled by a forced draught, and the novel use of a single mechanically driven valve, pressurised lubrication, rack and pinion steering. He also made innovations in suspension and transmission, and his was the first car which was designed specifically to use pneumatic tyres.

This was also probably the first four-wheeled petrol-driven car built in England, and the Automobile Club of Great Britain and Ireland in 1899 awarded him a gold medal in recognition of its many technical innovations. It also took part in the 1,000-mile trial of 1900. He further developed the car, prior to it being manufactured from 1901. However, Lanchester was no businessman and by 1904/5 the company was in the hands of the Receiver. It was rescued by Charles Pugh, but at the cost of Lanchester relinquishing ownership and managerial control and becoming a salaried designer on £250 per annum. Relations were never good and Lanchester resigned in 1914, later working for Daimler and Wolseley.

But Lanchester's brilliance was to have remarkably little immediate impact on the direction of car design. Pomeroy offers us three possible reasons for this. Firstly, some of his ideas were so far ahead of their time that few people could understand them, let alone apply them. In a similar vein, Lanchester's books on aerodynamics were brilliant but almost impossible even for specialists to comprehend (it is claimed that Lanchester took up an interest in the motor car after being warned that continuing his work on flying machines would ruin his reputation as a 'sane engineer'). Secondly, he was a renaissance man, and was not single-mindedly dedicated to perfecting the motor car. In addition to his long-standing interest in flight, he was also a potter of such skill that Bond Street dealers would offer to buy his works as genuine Chinese antiques. He also built aeroplanes, radios and gramophones, wrote works on acoustics and relativity, produced music and poetry and patented further important car engine developments in later life. Thirdly, for all his engineering genius, he was not able to escape from the preconceptions of his day. He thought in terms of horse-drawn vehicles and, having built one that performed the same function at twice the speed, he tended to think his work was done.

Even so, King claims that eighteen of the thirty-six primary features of the modern car were introduced by Lanchester. It is also worth noting that the owners of Lanchesters included Henry Ford, and that Ford's approach to the interchangeability of parts, making possible mass production, were anticipated in Lanchester's work.

HARRY JOHN LAWSON (1852-1925)

The introduction of the Horseless Carriage opens up such a wide vista of future utility that it seems impossible to foretell its illimitable prospects. Its effects upon everyday industries, upon mercantile progress, and upon passenger facilities for transit will unquestionably be enormous, and will mark the new departure as one of the most marvellous features of reform which the future historian will have to record in connection with an age unprecedented in its wealth of discovery and scientific research. It is doubtful whether in the whole brilliant record of useful inventions of which the nineteenth century has already proved so prolific there is likely to be affected a more wonderful revolution than that foreshadowed by the introduction of the Horseless Carriage. Not even the triumphs of Stephenson's ingenuity and genius, great and far-reaching as they have proved to be, are likely to be regarded as of more importance to the community at large than the new method of locomotion of which Mr Harry J. Lawson is the brilliant pioneer. The effect produced in this country by recent practical demonstrations of its utility has been little short of magical... Mr Harry J. Lawson to whose inventive skill and remarkable energy the public is indebted for this new form of locomotive power... Mr Lawson had already written his name largely on the pages of the history of industrial development... Mr Lawson's persistent labours and his confidence in the brilliant future of his new motor are at the present moment the talk of the commercial world...etc., etc.

(A small extract from a rather enthusiastic advertisement for the Horseless Carriage Company, and for Mr Harry J. Lawson, in The Times, 21 May 1896)

Harry J. Lawson was the most extraordinary figure in the early history of the English car industry. He has been described as the father of the British motor industry, though, with a parent like him, the infant industry should have been taken into care. His life in a number of respects paralleled that of George Hudson, the so-called Railway King, whose meteoric rise and fall was at the heart of the early growth of the British railway network.

Lawson was the son of a well-known Puritan preacher from Brighton. He was apprenticed as an engineer in 1868 and initially made money from inventions to improve the bicycle. His 1876 patented safety bicycle was said to be built from parts of the perambulator he had occupied as an infant. He was also involved in the establishment of the Rudge and Humber Bicycle Companies. In 1880 he took out a patent for what has variously been described as the first car, the first motorcycle and something looking like 'the result of an accident between a cable car and a perambulator'. Certainly, the scope of his Patent No.3913 was cast widely enough, referring as it did to 'Improvements in Velocipedes and in the Application of Motive Power thereto, such Improvements also being Applicable to Tram Cars, Traction Engines and other Road Locomotives'. On this shaky foundation Lawson claimed himself to be 'the brilliant pioneer' of Britain's, if not the world's, motor industry.

But Lawson was not an engineer or a manufacturer – he was above all a promoter. This was a great era for company promotion, offering a new class of prosperous but naïve investors the promise of getting rich quick on the back of new technologies, or by opening up exotic corners of the world for exploitation. The way it worked was that the promoter would buy

Harry Lawson (in his 'Swiss Admiral's outfit') and his wife, on the day of the Emancipation Run. The bearded man beside the car is Lawson's driver, Otto Mayer.

control of a business that looked promising but was under-capitalised. He would then launch it as a public limited company on the back of a wildly optimistic prospectus, with the aim of drawing in vast amounts of small investors' money. Having done so, and having paid back the cost of the company's acquisition, how much the promoter then invested back into developing the company, and how much he would pocket as straight profit, was a matter between him and his conscience.

Suffice it to say that Lawson's companies generally ended up with share prices vastly higher than the value of the company's assets, and with little or no prospect of paying a genuine dividend. If it was not actually illegal, it was distinctly sleazy. During 1894-95 Lawson had made a lot of money from converting cycle and tyre firms into public companies, and he foresaw the British car industry growing like its continental counterparts once it was deregulated. In 1895 he and two others set up the British Motor Syndicate Ltd (BMS). One of his partners in the venture, Herbert Duncan, explained the idea behind it. They were going to:

> Corner the British side of the motor industry by buying up all past, present and future patents in the expectation of running across some 'master' patents. They figured that the fees secured from royalties and manufacturing and selling licences would cause a continual flow of dividends and eventually raise the price of shares beyond the imagination of the most sanguine holder.
>
> (King, page 13)

The BMS was launched with a positive tidal wave of hysterical prose and a board of directors that gave the enterprise every appearance of respectability. It included nobility (the Earl of

Winchelsea and Nottingham), a motoring pioneer (Gottlieb Daimler) and the Count De Dion, who fitted into both categories. There was also one of the first pioneers of British motoring, the Honourable Evelyn Ellis, and Herbert Mulliner from the long-established coachbuilding company. Part of the sales pitch, which places Lawson somewhere between Leonardo de Vinci and God, is quoted by *The Times* above.

The rather speculative nature of this venture was not helped by the fact that Lawson appeared to have little grasp of the future direction of the evolution of the motor car, and even less of the value of money. It was said that 'he paid away £10,000 as the ordinary man would a £5 note.' BMS was set up with a capital of £150,000. They obtained an option on the old Coventry Cotton Mills, on part of which a factory had been built, apparently as a front for persuading people that he was intent upon manufacturing his own products. He also acquired the Daimler Motor Company's patent rights for England at a cost of £100,000, and personally received a payment of £40,000 for licensing them to build cars, to be built in his Coventry premises. He then set about acquiring the patent rights for De Dion Bouton, paying half of the £20,000 cost in BMS shares that would later prove to be worthless.

But Lawson's worst purchase was that of seventy or eighty patents registered by an American, one Edward Joel Pennington. Pennington had fled the United States, one step ahead of the exposure of a series of frauds, and was introduced to Lawson, who was persuaded to part with £100,000 (cash, not BMS shares, this time) for his patents. Pennington set up luxurious offices in the Coventry Mills where he conducted a high-rolling lifestyle and employed an unnecessarily well-staffed drawing office, no doubt in hopes of fooling other British investors. The man was, however, less of a manufacturer than a patent junkie, filing no less than twenty-eight applications in 1896 alone.

He had a long track-record in fraud. In 1890 his Mount Carmel Aeronautic Navigation Company held out the prospect of an all-aluminium airship, 200ft long and carrying forty passengers at 250mph. Its only shortcoming was that it could not leave the ground. Then, in 1895, he had introduced an early motorcycle, though some journalists of the day reckoned it was unlikely to run, let alone perform the elaborate stunts that Pennington claimed for it, asserting that '…the Pennington car … is not a car, since it does not carry, but has to be carried…' (*The Horseless Age*). Among its unique design features, the cranks of the engine were connected directly to the axle, and the engine had no cooling fins. Just to demonstrate his credentials as a well-rounded human being, Pennington also deserted his wife and several children in Cincinnati, setting up home in London with a rich Detroit socialite.

By 1898 some of his American creditors had caught up with him and he was forced to return to the United States to defend himself in court. Two years later and Lawson himself was suing Pennington for profiting from patent rights that he had already sold to BMS. Meanwhile, the Humber Company, who had fallen under Lawson's control prior to 1900, built Pennington's cars in very small numbers, not one of which ever managed to find a private buyer. Back in the United States, he was arrested in Philadelphia in 1901, after his latest invention – an armour-plated machine-gun-equipped *war automobile* that he claimed was capable of up to 130mph – caused several runaway horses, killed almost half a dozen pedestrians and nearly collided with several street cars.

By 1901 Pennington's British company had gone into receivership, and from 1904 his movements became increasingly difficult to trace. He is, however, known to have died suddenly in 1911, and is thought to be buried in an unmarked grave in Springfield, Massachusetts – his trail as well covered in death as it had been in life, according to an American biographer. The mechanical nightmare that was the Pennington motor car is described in detail later in the book.

Meanwhile, Lawson's claims for his company were as lavish and unfounded as were Pennington's for his inventions. At the first company meeting in 1896 he announced a bumper dividend:

...the first in the horseless carriage industry, of 10 per cent, at the rate of 30 per cent per annum – an interim dividend which I hope before the end of the year will be considerably more...The public has begun, however slowly, to wake up to the fact that a horseless carriage is a necessity.

(King, page 14)

In his speech, he stressed the other benefits of a car-owning society, not least among them the employment of chauffeurs, since 'it would be quite out of character for a lady (say, attending an evening party) to drive her own motor car; that cannot be expected. She must, of course, have her motorman.' With someone like Pennington designing the car, a driver with mechanical aptitude was certainly a must.

Lawson was active in every aspect of motoring. He bought the French designer Camille Bollee across from Paris and paid him £20,000 for the rights to his motor tandem; he founded the Motor Car Club, which had its offices in a luxurious suite at 40 Holborn Viaduct, issuing lavish uniforms to its members; and he claimed credit for the 1896 legislation loosening the restrictions on the use of the car (though his rival Sir David Salomons was the real guiding force behind it, and had far more disinterested motives for liberalising the car laws than Lawson's plans to corner the market in car-related patents). Lawson also organised the London to Brighton run in celebration of the Locomotives on Highways Act; and he promoted one of the first motor car shows in Britain at the Imperial Institute, South Kensington (Salomons mounted a rival show at Olympia the same week). In the midst of all this activity, he floated another venture, the Great Horseless Carriage Company, with a share capital of £750,000 (only half of which was paid up). This new company did not even have rights over most of the Lawson patents, its main asset being Lawson's premises.

Engineering magazine, at least, had seen through him, and was vitriolic about his share offers: 'The credulity of the investing public seems inexhaustible; no bait is too coarse, no lure too ill-concealed, to attract victims...', buyers of shares believing that the '...vulgar self-assertion and business shrewdness must always go hand in hand...' They spoke of:

Harry Lawson leased a four-storey factory with 105,760sq.ft of floorspace from which to take over the British motor industry. This purports to be a full week's output from his British Daimler Motor Company, lined up outside.

'A cunningly worded prospectus with lying promises to tempt the cupidity of the ignorant public...' and described Lawson as '...the company-monger, with his blatant self-advertisement, slimy persuasiveness and cozening ways...'

(Quoted in Nicholson, volume 3, pages 417-8 and 467)

But they warned in vain. In the wake of the publicity surrounding the London to Brighton run, the value of the shares in his new company increased three-fold. After two years the Horseless Carriage Company was converted into the Motor Manufacturing Company Ltd, but it was so ineptly run that, by the middle of 1904, the company fell into the hands of the Receiver. In the meantime, he had separately launched a further share issue for the BMS. Throughout all this he continued to issue grotesquely large dividends, paid for in worthless Great Horseless Carriage Company shares. Lawson was by now a man not to be trusted by the business community. As the *Stock Exchange Gazette* put it, 'the fact that Harry J. Lawson is the controlling spirit is a very bad omen for the company and augurs a speedy acquaintance with the bankruptcy court.'

Lawson's lack of engineering expertise meant that he did nothing to inhibit the tendency of his directors' engineers to 'improve' the proven designs of the French cars to which he acquired the rights, generally with disastrous results. His patents also became increasingly difficult to uphold in law. In defiance of Lawson's monopoly, S.F. Edge put a Napier engine in a racing Panhard originally owned by Lawson, from which emerged the Napier that performed so well in the 1,000-mile reliability trial of 1900. The Count de Dion was able to force Lawson to concede him a separate agency to sell his own products throughout the British Empire. In 1904 Lawson was indicted for conspiracy to defraud and for making false statements to would-be investors. He conducted his own defence and was crucified by the press, before being sent down for a year's hard labour. What remained of his portfolio of patents was bought for just £1,000 by Napier in 1907. Lawson's business career was in tatters and he died, almost penniless, in 1925.

LOUIS PANHARD (1841-1908)

Panhard et Levassor were, with Peugeot, the world's first car manufacturers. Louis Panhard was the son of an entrepreneur with coachbuilding and rent-a-carriage businesses that dated back to 1830. They also went into the manufacture of woodworking machinery, trading under the name of Perin, and in 1853 patented a bandsaw. During the Franco-Prussian war of 1870-71 they diversified further into the manufacture and boring of cannon, and it was this that led them into metal-working during a down-turn in their other businesses. They began making Deutz gas engines under licence and, in 1872, a friend of the family, Emile Levassor, joined the business. It was his marriage to Madame Sarazin, the holder of the French Daimler patent rights, which brought them into the motor car business. They produced four identical cars during 1891, which, along with Peugeot, constituted the world's first production run (Benz and Daimler pre-dated them with their prototypes, but had not at that time gone into production).

ARMAND PEUGEOT (1849-1915)

Peugeot were an established cycle manufacturing company well before the age of the motor car. Their first involvement with the horseless carriage was building a number of chassis for Serpollet steam engines in 1888/89. They showed one at the 1889 Grande Exposition in Paris. Gottlieb Daimler and Emile Levassor had already seen examples of Peugeot vehicles, and persuaded Peugeot to build some cars, under a Daimler licence and to their design, using a Panhard et Levassor engine. The three-way contract was signed and the first prototype was delivered in April 1890. They went into production at their factory at Valentigney in 1891.

Although they only produced five cars that year, all of which differed in some details, this makes them, along with Panhard et Levassor, the oldest car manufacturers in the world.

Their first models had in common a twin-cylinder rear-mounted engine, chain drive, a tubular steel chassis and tiller steering. Their first model could not be described as totally developed since it loaded all the weight of the engine and the passengers onto the rear wheels, leaving the steered front wheels dangerously light. It also featured the blind alley of a vis-à-vis seating arrangement, in which the driver sat in one of the rear seats and tried to peer around the passengers seated in front, facing him. By 1895 they had introduced a front-engined model (by which they meant an engine under the front seat, since the vehicle still had a four-seater vis-à-vis arrangement). It was at this point that Armand Peugeot decided that the licence from Daimler was proving to be an unreasonable restraint on his overseas trade, and began manufacturing his own motors. More radically, two years later he parted company with the family business and set up his own company, S.A. des Automobiles Peugeot, with a new purpose-built factory at Audincourt and a head office in Paris. For a while, from 1905, there were two Peugeot companies producing cars (the old factory made the Lion Peugeot – the trademark lion motif coming from the coat of arms of the town of Belfort) until, in 1910, the two firms were re-united. The company had some success in Grand Prix before the First World War, and won the 1913 Indianapolis 500. Their racing engine of that period was to form the basis of the American Miller and Offenhauser engines. Their other notable development in this period was the production from 1912 of an early model by the legendary designer Ettore Bugatti – the bebe Peugeot.

This extraordinary Lion Peugeot voiturette racer dates from 1909. From 1905 there were two manufacturers of Peugeot cars, after Pierre and Robert Peugeot bought themselves out of an agreement signed by their father, so as not to build cars in competition with his brother Armand. The companies later re-merged.

In an attempt to limit the ever-growing speed of racing cars, a limit was put on the bore of the cylinders (or cylinder). In voiturette racing, this led to peculiar long-stroke engines like the one in this 1909 Lion Peugeot. The bizarre silencer arrangement was apparently added to overcome objections about noise when the car ran unsilenced at Brooklands.

LOUIS RENAULT (1877–1944)

Louis Renault was the third son of a cloth merchant, who was introduced to motoring at an early age with a ride on one of Serpollet's steam cars. In 1898, aged just twenty-one, he built his own first car in the gardener's cottage of the family estate. His aim was to make a car that would be something between the De Dion tricycle that he already drove and a 'proper car', and he did so in just three months. It was a front-engined model with the innovation that it had shaft drive. He used an off-the-shelf De Dion motor and as many pre-assembled parts from the cycle industry as were suitable, in order to keep the costs down. The original intention had been to sell on the manufacturing rights, but people's reaction to it was so positive – after Louis demonstrated that it could climb the slope of Lepic Street in Montmartre – that he, along with his brothers Fernand and Marcel, decided to build it themselves. The Renault Company was launched on 21 March 1899. The company took off quickly; by the end of the year they had sold sixty cars, had introduced a new water-cooled De Dion engine, had taken on 100 new staff and doubled the size of their workshop. All of this was underpinned by Renault's competitive success, winning the Paris–Trouville and Paris–Ostend races.

The power and speed of his machines grew rapidly. His first model was just 240cc (some sources say 198 or 270cc) and 1.75hp, but within two years he had more than quadrupled the size and horsepower to 1,012 and 8 respectively. In the Paris–Berlin race of 1901 he finished eighth overall, beating all the other voiturettes and light cars. The following year he stepped up further when he acquired the rights to a four-cylinder engine designed by M. Viet (the brother-in-law of Georges Bouton, whom Renault poached from De Dion Bouton). With this, his brother Marcel entered the Paris–Vienna race of 1902 where, as we saw, his 3.8-litre

Louis Renault, pictured in his voiturette in 1899.

24hp entry in the light car class beat the 13.7-litre 70hp monsters from the likes of Panhard to win the overall prize, to prove for the first time that science (aided by some rather reckless driving, it was alleged) could triumph over brute force.

In the following year's Paris–Madrid race, Louis was in second place when the race was abandoned at Bordeaux. One indication of the dangers of that road race was the fact that he reached Bordeaux at an average of 62mph. Of more immediate impact was the fact that Marcel Renault was killed during the first stage of the race. Renault immediately announced their retirement from racing. The decision was not long-lasting, and Louis Renault's mechanic, the Hungarian Francois Szisz, won the first French Grand Prix in 1906 on the Sarth circuit, with the consequence that Renault's sales rose from just over 2,000 in 1905 to 4,000 in 1908, almost 7,000 by 1910 and over 10,000 by 1913.

CHARLES STEWART ROLLS (1877-1910) AND HENRY ROYCE (1863-1933)

Frederick Henry Royce, the technical brains behind the famous partnership, trained in railway engineering before setting up his own manufacturing company in Manchester in 1884. They produced everything from electrical switches to large pieces of engineering such as cranes. He bought a French Decauville car, and its imperfections led to his interest in manufacturing an improved product. From tinkering with his Decauville, he rapidly moved to designing and building his own car. This prototype was seen by the Hon. Charles Rolls, a wealthy young man who was already running a profitable car sales business, who was so impressed that, within months, he had given up selling other makes and become a partner in Rolls-Royce. There was little by way of innovation in the cars they produced; they simply incorporated all the elements of best practice from other makes, superbly executed. Royce built two- and three-cylinder cars in the early years, but his masterpiece was the six-cylinder, (initially 7-litre) Silver Ghost, launched in 1906, and which continued in production (with constant development) until 1926, becoming the showpiece of motor engineering from that period. Rolls-Royce continued the coachbuilding tradition, supplying chassis for others to fit with bodies.

Not everyone was a fan of Royce's approach. Laurence Pomeroy said that Royce produced the pinnacle of nineteenth-century engineering in the early part of the twentieth century, whereas Lanchester designed the first scientific twentieth-century car in the latter part of the nineteenth century. He added for good measure that 'the Rolls-Royce is a triumph of workmanship over design'. It was certainly the case that Royce's perfectionism created problems for the company. He eventually had to be virtually barred from the Derby factory for trying to set standards that would have rendered the company hopelessly uneconomic.

CHAPTER 8

THE BRITISH
EXPERIENCE

'The new French sporting craze'
(How much of Victorian Britain saw motoring)

In this chapter we look in more detail at how motoring emerged in Britain. The story begins with the punitive legislation against the motor car and the battle to liberalise it. Closely related to this battle is the slow start to the British motor industry and the ambivalent attitude of the public towards motorists.

EARLY MOTORISTS AND THE LAW

Discrimination against self-propelled vehicles had its origins long before the age of the internal combustion engine. As we saw earlier, once the turnpike trusts became aware of the potential of steam carriages many of them began subjecting steam transport to punitive tolls. Many reasons (real or fabricated) were given for this. One was the fear that the steam carriage would drive the turnpike's main customers – the stagecoaches – out of business. There were also (unproven) claims that these new fire-breathing monsters on the turnpike would make all the horses bolt, and that their weight and speed would do untold damage to the road surfaces. Then there were those, like the unemployed handloom weavers, whose bitter experience of mechanisation led them to hate machinery of all kinds, and who physically attacked early steam carriages.

Attempts were made to get the tolls for steam carriages put on the same footing as stagecoaches. The 1831 Bill went before a Commons Select Committee, where it was told by no less an authority than Thomas Telford (among others) that steam carriages did far less damage to the road than stagecoaches. In fact, all the evidence put before the committee supported steam transport, and this was reflected in their report, which concluded that:

As this 1901 *Punch* cartoon entitled *Brothers in Adversity* shows, the horse and the early motorist enjoyed a somewhat fraught relationship. he original caption reads: 'Farmer: "Pull up, you fool! The mare's bolting!" Motorist: "So's the car!"'

> These enquiries have led the Committee to believe that the substitution of inanimate for animal power, in draught on common roads, is one of the most important improvements in the means of internal communication ever introduced.
>
> (Quoted in Nicholson, page 90)

In March 1832 a further Bill was introduced to give steam carriages parity with the tolls charged to stagecoaches for a period of five years. It cleared the Commons, and then disappeared into a Lords Select Committee. Once again most of the people giving evidence to it were known to be pro-steam; but none of its evidence was ever published, and the Bill was mysteriously thrown out. Its loss also signalled the disappearance of most of the potential investors in steam carriages. In that same parliamentary session, fifty-two new turnpike Acts were passed. Of these, thirty-nine had separate tolls for steam carriages, almost all of which were discriminatory, charging between two and eight times the toll for a stagecoach.

A further attempt to get the tolls reformed went to the Commons in 1836. Unfortunately, as we saw in the period 1831–34, a number of steam carriages had suffered boiler explosions – not the most promising of contexts for promoting the cause of steam. Once again the Bill got as far as a Lords Select Committee, although this time its deliberations at least got into the public realm. The Bill was once more rejected, but it mattered little by then. Interest in steam carriages was waning as the burgeoning railways sucked in all the available investment capital.

It was the railways, and not steam-driven road vehicles, that did the real damage to the turnpikes through competition. Many turnpike trusts went into liquidation, leaving the under-funded parishes and their press-ganged labour forces to maintain the roads, most of which turned back into deserted country lanes. The inns along them closed and the gate-keepers were left to find such alternative employment as they could. But whilst steam passenger vehicles came to nothing, the lumbering steam traction engine became a relatively common sight on the nation's roads, with an estimated 8,000 of them in use in Britain by the 1890s.

> If I had my way, I would make it a crime to use automobiles on the public highways....
> Perhaps the time will come when horses will be educated to the point where they will not be afraid of automobiles; but I doubt that, for I have not yet seen the time when I was not afraid of them myself.
>
> (Senator Joseph W. Bailey of Texas, quoted in Setright, pages 11-12)

But prejudice against motorised vehicles was not limited to Britain. Benz was told in 1893 that the authorities in Baden were going to introduce a speed limit of 7.5mph on country roads and half that in towns and on sharp corners (however these were to be defined). In 1899 the authorities in Mannheim banned Benz (again, rather impractically) from driving on streets of less than 11.3m in width. In New York laws similar to the British Red Flag legislation applied until 1901. The Prussians tried to ban steam cars under an obscure law outlawing 'occupied premises above a boiler.' Switzerland had a total ban on Sunday motoring until 1927, and one canton, Graubunden, banned motoring entirely from 1900 to 1925. The police in Nice carried swords with which to prick the tyres of any offending motorist. In Milan you had to give the authorities details – date, time and route – of your proposed trips, while the American state of Tennessee demanded a full week's notice of any intended car journey. Other American legislation required motorists to cover their car with a cloth if it frightened an approaching horse. Pettifer reports a proposed American Bill that would have required any self-propelling vehicle to come to a complete halt on approaching any crossroad, which might have been the start of an extraordinary process:

> ...the engineer must thoroughly examine the roadway ahead and sound his horn vigorously. Then halloo loudly or ring a gong... discharge a Roman candle, Vesuvius bomb or some other explosive device as a final warning of his approach.
>
> (Pettifer, page 11)

Meanwhile, legislation that might actually have been useful – such as the introduction of a driving test – took thirty years of lobbying before it was enacted in Britain in 1935. Despite the absence of a test, there was no shortage of would-be British motorists queuing up to pay three guineas for a course of twelve driving lessons. The French brought a compulsory driving test in during 1910. In other respects, though, they tended to have a more liberal attitude towards motoring and motorists, though even they were challenged by a lady who appeared on the streets of Paris in 1901 in a vehicle pulled by two cows, demanding that her conveyance be licensed as a horseless carriage.

THE 'RED FLAG' ACTS AND THEIR REPEAL

The real impediment to the motor car lay in legislation passed in the middle of the nineteenth century. The Locomotive Act of 1861 (24 and 25 Vic. cap. 70) set speed limits of 10mph in the country and 5mph in towns for 'road locomotives'. But it was the so-called 'Red Flag' Act, the Locomotives Act of 1865 (28 and 29 Vic. cap. 83) that effectively crippled the development of mechanical road transport in this country for the next thirty years. It reduced the speed limit for powered vehicles to 4mph in the country and 2mph in the town, and required three people to operate such vehicles. One person was to walk sixty yards in front of the vehicle carrying a red flag to warn oncoming traffic of its approach; the other two were to drive and (rather superfluously, in the case of the internal combustion engine of the 1890s) to stoke the vehicle.

It also gave local authorities powers to restrict the hours of operation of locomotives, to charge a licence fee of £10 for every local authority area in which they operated, and to forbid locomotives to cross bridges unless the bridge bore a notice permitting it. Thomas Aveling, a traction engine manufacturer, summed up the injustice of this last provision as follows:

> If I send a boiler weighing fifteen tons and drawn by fifteen horses over a bridge and that bridge breaks, I have nothing to pay. However, if I send that same load drawn by a traction engine over the same bridge, and it then breaks, I am fully responsible for the damage.
>
> (Wilkes, page 45)

The same prejudice was to be found on toll bridges, which might charge 3*d* a ton for a horse-drawn load, but 4*s* a ton for the same load drawn by a traction engine. One exception to this generally restrictive regime was that, from 1879, street trams were authorised to use mechanical traction under the terms of licences granted by the Board of Trade, though this again was subject to conditions which included speed limits.

Supporters of steam traction tried again to get a Bill through Parliament in 1873 relaxing some of these regulations, but it fell in the face of determined opposition in the House of Commons. What legislation as did get through, such as the Highways and Locomotives (Amendment) Act of 1878 (41 and 42 Vic. cap. 77), simply tinkered with details such as weight limits and the construction of wheels without addressing more fundamental problems. It is, however, worth noting that it was this Act, rather than the one of 1896, that removed the need for a man out front with a red flag. He still had to walk in front, but he no longer had to carry a flag, and now only had to be twenty yards in advance of the vehicle. But, at the same time, this Act gave local authorities the power to ban the operation of such vehicles entirely for eight hours out of every twenty-four, so it is debatable whether, overall, it represented any advance at all. There also continued to be a widely held misconception that the red flag requirement remained in force.

As a result, the British car industry continued to be severely stifled for the next decade or more. Stifled – but not to the point of total extinction. A few electric buses and goods delivery vehicles ran in London and elsewhere. Charles Gerrard drove an electric phaeton

1896 and the Hon. Charles Rolls in his Peugeot is shown being preceded by a man with a red flag.

from Coventry to Birmingham at an illegal 10mph in June 1894, and there is even a record of a prototype British-made petrol car almost coming into conflict with the law in the same year. One Frederick Bremer built a crude copy of a Benz (to give an indication of how crude, it had a grindstone for a flywheel and sparking plug insulation made from a clay pipe stem). A mounted policeman attempted to detain him, but the noise of the vehicle caused the policeman's horse to bolt and Bremer escaped prosecution.

But the immediate future for motoring in Britain lay with imported vehicles. The first petrol-engined car was imported to Britain in November 1894 by a man called Harry Hewetson. He set out deliberately to flout the 1865 Act and make a mockery of it, in order to highlight its absurdity. He had a small boy walking in front of the vehicle, waving a red flag the size of a postage stamp, attached to a pencil (since the regulations – even if they had still been in force – did not specify what size of flag was required). The Paris–Rouen event, which took place the same year, was also important in starting to mobilise public opinion against the Act, and others began to follow his example. The following year the Hon. Evelyn Ellis bought the Panhard he had previously kept for use in France to England, and drove it, without the assistance of a red flag of any description, on a fifty-six-mile journey at an average of 10mph with the express intention of attracting a summons. However, despite press coverage of the trip (which reported that just two of the 133 horses they had passed along the route had shied) the police seemed to be singularly unwilling to enforce the law. They were equally reticent when Ellis followed his trip by an even longer one of 120 miles, from Datchet to West Malvern. Another pioneering car builder, John Knight, was less fortunate, attracting a speeding ticket for doing 8mph in his home town of Farnham in November 1895.

Meanwhile, the campaign against such restrictive legislation was being led by *The Engineer* magazine, which complained that:

> The petroleum carriages can go as fast as an ordinary cyclist cares to travel…The authorities in the United Kingdom appear to do what they can to hinder the use of such carriages, whereas in France…everything possible seems to be done to encourage their use.
>
> (*The Engineer*, March 1894)

By now Parliament had changed since the passage of the 1865 Act. Its membership had become more urbanised, with relatively fewer representatives of agricultural interests (many of whom had a vested interest in maintaining the supremacy of the horse, being suppliers of horses or horse food) and more members from a financial or commercial background. Ironically, however, it was because of agriculture that the Local Government Board, which controlled much of the regulation of motor vehicles, stated that:

> Mr Mcdona to draw the Local Government Board's attention to the fact that the winning car from the Paris-Bordeaux race subsequently covered the 56 miles from Micheldever (Sussex) to Datchet (Berkshire) in 5½ hours, carrying four people at a petrol cost of ½ a penny an hour. He asked them 'whether, in view of the fact that this new and cheap motive power might be made use of by farmers and others to bring their produce to market, the right honourable gentlemen will consider the advisability, in the interest of the depressed agricultural industry, of repealing the Locomotive Act of 1865, so to admit of petroleum and electric motors being more extensively used in the United Kingdom'.
>
> (From a parliamentary report in *The Times*, 24 August 1895)

Agriculture was in recession, and the case was made that, by its ability to effect door-to-door collection and delivery of agricultural produce quickly and cheaply, the motor vehicle could be the salvation of farming. The argument was taken up with enthusiasm: 'the motor car is the key which will solve the problem of agricultural depression…the best friend of the farmer will be the motor car'.

An anti-horse case was also made for the car, on the grounds that the amount of manure that horses dropped on urban streets every year was a health hazard that would be removed once they were replaced by the motor car. There was substance to this case. There were around 300,000 horses working on the streets of London in the 1890s and, in America, someone calculated that 1.3 million pounds of horse manure was deposited daily on the streets of New York. It is hard for modern readers to appreciate quite how unpleasant an environment this created. As one contemporary account put it, the streets were:

> ...literally carpeted with a warm, brown matting of comminuted horse dropping, smelling to heaven and destined in no inconsiderable part to be scattered in fine dust in all directions, laden with countless millions of disease-breeding germs.
>
> (Bak, pages 18-19)

Moreover:

> ...if, while crossing the street, you had the misfortune to get too close to a passing cab, you would emerge looking like a Dalmatian dog, spotted with thick brown viscerous (sic) mud (much of it simply horse dung).
>
> (Thorold, page 3)

Horse manure provided an ideal medium for the transmission of one of the major killers of the day, tuberculosis. It also encouraged flies bearing other diseases and was generally said to reduce life expectancy by five years. Moreover, dead horses were often just left in the street, and the New York authorities were said to remove 15,000 decomposing carcasses a year. The incidence of disease did indeed fall as the number of horses (living and dead) on the streets reduced, and it is difficult to comprehend from today's perspective that one of the first claims for the motor car was that it brought clean air to our towns and cities. This view was underwritten by the 1898 Conference of Municipal Engineers, who resolved that:

> ...the introduction and use of efficient motor vehicles should be encouraged by county, municipal, urban and other authorities, in view of the fact that the extended use of such vehicles would contribute to the general improvement of the sanitary condition of our streets and towns.
>
> (Setright, page 15)

Thorold has an interesting section dealing with one further and unfamiliar claim made on behalf of the motor car for its effect as a tonic for the digestion and the nerves, which came from an eighty-one-year-old doctor, Sir Henry Thompson, in 1901:

> The easy jolting which occurs when a motor car is driven at a fair speed over the highway conduces to a healthy agitation. It 'acts on the liver' to use a popular phrase, which means only that it aids the peristaltic movements of the bowels and the performance of their functions...

He goes on to recommend car travel as a cure for insomnia, and reports that he has known:

> ...instances of ladies suffering from defective nerve power who have derived great benefit from the invigorating and refreshing effect of meeting a current of air caused by driving in an automobile.
>
> (Thorold, page 24)

A German article thought the benefits might be as much spiritual as physical, speaking of:

…a highly advantageous effect on the nerves (that) goes hand in hand with the beneficial relaxation caused by scenic landscapes and the unburdening of the internal organs.

(Ibid)

Others offered quite the opposite diagnosis, very much reflecting the views of a Mr Toad, who we shall encounter shortly:

Motoring, it must be admitted, is a disease, a sickness of the mind. And it has a very pretty name – la vitesse… (the motorist) cannot keep still – quivering nerves tight as springs, impatient to be off again the moment he arrives, in agony to be somewhere – anywhere – else.

(Octave Mirbeau, French novelist and motorist, writing in 1906,
quoted in Thorold, page 25)

The horse-ridden city streets were also dangerous and congested places. Traffic jams were in many cases as bad as in modern cities, and there were as many people killed by horse-drawn vehicles in 1872 as there were people killed by motor vehicles a century later. Also on the charge-sheet for the horse was the fact that it diverted large swathes of agricultural land away from human food production (an argument used against bio-fuels for motor cars a century later). But this did not stop the horse lobby campaigning against the motor car:

Motorists have compelled one hundred thousand people to withdraw their horses and carriages from the public roads. It is estimated that one hundred thousand men have been thrown out of work as a consequence…

(Handbill circulating in London, 1908, quoted in Pettifer, page 169)

The motor car soon found itself a number of influential supporters in Parliament. It was at this time that two very different – and bitterly opposed – champions of the motor industry emerged. One was the disreputable but flamboyant entrepreneur Harry Lawson, whose dubious dealings are described elsewhere in the book. The other was the much more principled, but rather more reticent, Sir David Salomons. It was Salomons who organised the United Kingdom's first motor exhibition at the showground in Tunbridge Wells (the town of which he was at one time mayor); who produced a pamphlet promoting the merits of the horseless carriage and who did much lobbying behind the scenes to win over key opinion formers, both in Parliament and elsewhere. Salomons also founded *The Motor* magazine, and prior to that established the Self-Propelled Traffic Association to promote the cause of 'automobilism'.

Lawson originally joined the association, but resigned to found the rival Motor Car Club, which pretended independence in pursuing 'the protection, encouragement and development of the motor car industry', but which was much more a vehicle for the promotion of Lawson's business empire. Given Lawson's apparently dominant position in the industry, he managed to enlist the support of many of its pioneers including, from its launch in November 1895, the editorial voice of the *Autocar* magazine. The first editor of this publication, Henry Sturmey (he of the John o' Groats to Land's End trip referred to in chapter one), would later be sacked for using the magazine as a mouthpiece for Lawson's propaganda.

It was Salomons who advanced the argument that, under the terms of the Customs and Inland Revenue Act of 1888, a motor car owner who took out a carriage licence should be allowed to exercise the rights of a carriage user. This included exemption from the provisions of the Locomotives on Roads Act. This case was much argued in courts up and down the country as motor cars became more numerous (or, at least, better-known), and prosecutions began to take place. However, it was not one that the courts would accept, though some did at least impose only nominal fines, and those with regret. This is not to say that cars became commonplace. Nicholson suggests that, by the end of 1895, there were probably no more than about two dozen motor cars on British roads, half of them home-built one-offs and the remainder imports. No reliable official records of numbers existed until 1904, when the registration of cars was introduced and a total of 8,645 private cars were recorded in England. This had risen to 72,000 by 1911 and 132,000 by 1914.

It was the Locomotives on Highways Act of 1896 (59 and 60 Vic. cap.36) that removed the worst of the speed restrictions relating to road vehicles of up to three tons unladen weight. This permitted a maximum speed of 14mph, but gave the Local Government Board powers to reduce it (which they duly did, lowering it to 12mph with effect from the day of its enactment, 14 November 1896).

As the Bill completed its various stages through Parliament, all sorts of arguments were advanced against the motor car. Her Majesty's Chief Inspector of Explosives was concerned – perhaps understandably – about the dangers of motorists storing large amounts of highly inflammable fuel in their homes (this in the days before filling stations). The Irish Nationalists also opposed the Bill as part of their attempt to wreck other legislation in pursuit of Home Rule. They advanced arguments which were a combination of the plausible (fears for the future of the Irish horse-breeding industry), the short-sighted (that cars were currently expensive and unreliable) to the positively bizarre (that the motor car was an insidious Jewish invention – possibly a reference to it being championed by Sir David Salomons who, among other things, eventually became the first Jewish Lord Mayor of London). There were also attempts to de-rail the proposals by giving the Local Government Board blanket powers to regulate or even prohibit motor traffic.

The Bill finally received royal assent on 14 August 1896, to come into force three months later. Its main provisions can be summarised as follows:

- They applied to vehicles of less than three tons unladen weight (four if pulling a trailer). The weight of fuel, batteries and water was not to be included
- The vehicles should not emit smoke or vapour
- The vehicles should be regarded in law as carriages
- The vehicles should carry a bell or some audible warning of approach, and lights (to indicate their presence to other road users rather than to light the way for the driver)
- The drivers should adhere to a maximum speed limit of 14mph, or such speed as the Local Government Board should specify
- Separate regulations should be laid on the storage and use of petroleum
- The Local Government Board have powers to regulate the construction and use of motor vehicles, and local authorities to be able to request such regulations in the interests of safety
- County councils have powers to regulate the use of motor vehicles on their bridges
- There is a scale of excise duties for light locomotives and fines for breaches of these regulations.

In due course the Local Government Board and the county councils produced regulations to put flesh on the bones of the Act. Some of these were meaningless tokens, specifying for example that drivers had to be competent and vehicles soundly constructed, when no provision was made for testing either (Salomons was apparently responsible for getting the Local Government Board to drop proposals for testing). Others were more restrictive. The 14mph speed limit was reduced to a maximum of 12mph (or no greater speed than was regarded as 'reasonable and proper' – whatever that meant – by the county councils). Even this speed limit only applied to vehicles under one and a half tons, while those of one and a half to two tons were limited to 8mph and those over two tons to just 5mph. The 12mph figure equated to the limit applied (not always rigorously) in Paris, and also to what was regarded as a normal speed for cyclists. They also added to the Act in a number of respects. A car weighing over five-hundredweight had to be fitted with a reverse gear, and there were minimum tyre widths ranging from 2.5in on vehicles weighing under a ton to 4in on those of two tons or more. There were separate Home Office regulations on the storage of petrol (discussed elsewhere).

Restrictive as these rules were, many still saw them as an unhappy compromise. In Scotland (over which the Local Government Board did not have control) they reduced the maximum speed further to 10mph. A number of English county councils (thinking they had the power to do so independently) tried to introduce unilateral blanket speed limits in their areas, some as low as 6mph, until they discovered that the Board had to ratify them and would refuse to do so. The Board thereafter found themselves playing pig in the middle between those seeking greater restrictions on the motor car and those who felt the law was still too restrictive. In the months leading up to the introduction of the new rules Lawson and others staged a variety of promotions to win over public opinion toward the motor car. In some cases they (unsuccessfully) sought dispensation from the police to run on the roads under anticipated regulations.

But generally the police maintained their hostile anti-car stance right up until the last minute. On 9 November 1896, just five days before the Act came into force, they applied for a summons against Lawson for driving up the Strand without having a man walking in front. He happened to be taking part in the Lord Mayor's procession at the time. The magistrate decided it was a purely technical offence and declined to issue the summons.

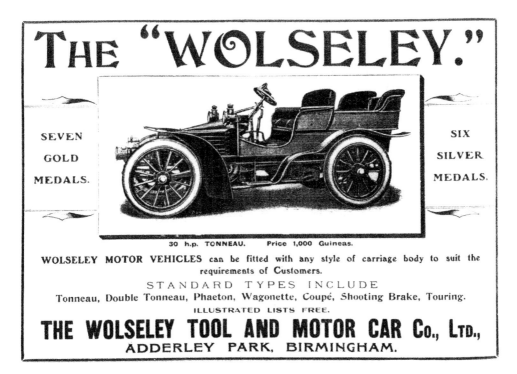

The young Herbert Austin cut his teeth on some of the earliest Wolseley models, which were the best-selling British cars before the First World War.

EMANCIPATION RUN 1896: LONDON TO BRIGHTON

> ...one must hope that the trial trip to Brighton will not be marred by any casualty,
> though it would be surprising if a first attempt such as this was exempt from mishaps of every
> kind.
>
> (*The Times*, 7 November 1896)

On the first day on which the new regulations took effect, Lawson's Motor Car Club duly
organised a celebratory 'procession' from London to Brighton (it could hardly be described as a
race, and certainly could not be publicised as such with the new speed limit of 12mph in force
on Britain's roads). The Bollee brothers were invited to take part and suffered the indignity of
being towed to the start behind horses. They arrived by train in London just before the 1896
Act came into effect, and decided to drive their vehicles from Victoria to Holborn where the
entrants for the run were being prepared. First they had to obtain benzene. There being no
petrol stations, the supplies of chemists within a wide radius were soon exhausted, and they
were ready to leave. But such was the noise and smoke when they started up their vehicles, and
such was the impact on a public totally unused to the internal combustion engine, that a police
constable was summoned. He told them it was not yet legal for them to drive on the roads, and
they were forced to hire three horses to tow the cars to their destination. This they did to hoots
of derision from passing cabbies and horse bus drivers, whose day was brightened further when
one of the cars' tow ropes broke in the middle of Piccadilly Circus, leaving it stranded.

Mindful of the hostile elements in the public's opinion, drivers were told on the day of the
run to stay in close order and to go slowly in procession for the first three-and-a-half-mile
stage of the journey to Brixton. Lawson's expensively purchased *Paris–Marseilles* Old No.5
Panhard (he paid £1,200 for it as a publicity stunt), then said to be the fastest car in the world,
was supposed to be the procession's pilot car, and was not to be passed by the others 'unless by
necessity'. The idea was that the participants would all reassemble on the outskirts of Brighton
for a further procession to the finish. Lawson, never given to understatement, dressed his Motor
Car Club members in garish uniforms. As the *Engineer* put it, it was 'not exactly the garb of a
German band, not exactly the dress of an excursion steamer steward, but something between
the two, only more pretentious.' Others thought the nautical theme had been combined with
that of a Hungarian bandit. Leon Bollee described them most succinctly as looking 'like Swiss
Admirals'.

The event undeniably drew crowds. They were lined nine or ten deep to watch the start, the
pavements were solidly packed for the first nine miles and, in total, it was estimated in *Autocar*
that some half-a-million people saw the event at one stage or other. Since many more wanted
to follow the event than could be accommodated in the cars, a special train was laid on for
spectators to travel from London to Brighton. Following a pre-run breakfast at the Metropole
Hotel (tickets 10s, including wine), the race (or procession) itself was started by the ceremonial
tearing up of a red flag.

Thirty-two, thirty-three, or possibly thirty-eight (the contemporary records are vague) of the
original fifty-eight entrants arrived at the start (at this time the Motor Car Club estimated that
there were something less than 100 motor cars in the whole of Britain). Harry Lawson owned
interests in thirty-four of the fifty-eight vehicles (and twenty-two of the starters) in one way
or another. The others he described as 'all kinds of experimental cars', as opposed to the 'tested
cars...lent by the British Motor Syndicate or their licensees'.

The entrants included the aforementioned Bollee brothers, an ailing Gottlieb Daimler and
Mayade, the victor of the Paris–Marseilles race. Entrants included electric cars, some of which
would have to do part of the journey by train (their range was only twenty miles). They made
for Victoria Station where they boarded a train bound for Preston Park just outside Brighton.
There, it is alleged, some of them muddied their cars to make it look as if they had travelled
to Brighton by road, and made for the finish. British interest was maintained by a number of

The Emancipation Run attracted huge crowds. Here they watch the participants (for we must not call them competitors) arriving in Reigate.

Harry Lawson's so-called British-built Daimlers (he had bought the manufacturing rights to Daimler engines, supposedly for use in his 'motor mills' in Coventry). However, there was precious little British content in these cars – they were French Panhard-Levassors and Peugeots with German-designed Daimler engines.

Several of the petrol cars would break down or suffer accidents en route, arriving in an undignified manner in Brighton, being towed by horses or carried on horse-drawn carts. But in other respects the event was well organised. Supplies of fuel and lubricating oil were laid on at the lunchtime stop at Reigate, and participants were given a list of places where water might be obtained en route. There was even one enterprising blacksmith at Albourne Green who advertised 'Motor cars repaired while you wait', though how much experience he had in the motor industry can only be guessed. The event proceeded at the planned sedate pace as far as Brixton, by which time Lawson's car was showing signs of overheating. The other contestants, in particular the overseas ones, to whom the idea of anything other than an out-and-out race was in every respect foreign, seized upon opportunity, skipping lunch, overtaking Lawson and setting off for Brighton at full racing speed. The average speeds finally recorded for the so-called procession by the leading finishers were on a par with the outright racing events of the day, such as the Paris–Marseilles race, which was won at an average of 15.7mph. The winning London to Brighton driver's completion of a fifty-two-mile journey in three hours and forty minutes (an average of around 14.2mph, on worse roads than those in France) suggests that the new 12mph speed limit was somewhat honoured in the breach, with some of the organising officials being among the worst offenders. The Bollee brothers rather confirmed this as they dodged in and out of Brighton's horse-drawn traffic in their racing tri-cars at speeds of up to 30mph. The effect on their fellow road users was no doubt enhanced by the racing Bollees' deafening exhausts.

Mercifully, by the standards of the day, the casualties were light. The only serious injury of the event was a small child who was run down in Crawley and suffered a fractured skull.

Among the more mysterious finishers were two American Dureyas, which some said had not been seen at the start of the race, nor en route, and it was suggested by some that they had been smuggled into Brighton the night before by train. This did not stop Dureya from exhibiting their cars as the victors of the event when they later toured with Barnum and Bailey's Circus, back in the United States. But, on the other hand, this accusation does not square with the fact that one of the Dureyas was also blamed for the running down of the child in Crawley. The procession was officially 'won' (insofar as one can win a procession) by Leon Bollee, with his brother Camille in second, twenty minutes behind. Others disputed the result, issuing sworn statements that the Dureyas had won and implying that the Bollees had cheated in some way. There were also counter-claims that Lawson did not allow the Dureyas to win since they were some of the few vehicles entered in which he had no commercial interest. The records were equally vague about how many cars finished, with claims of anything between twelve and twenty-two. Some, plagued by breakdowns, did not arrive until the following day.

However, the public response to the event seems to have been generally favourable, despite the general acceptance on their part that it had been an out-and-out race (something rather confirmed by the Motor Car Club awarding gold medals to the first eight finishers). *The Autocar*

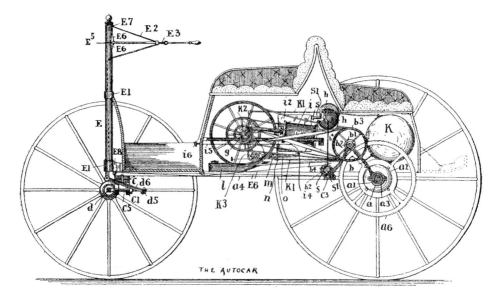

The mechanical mysteries of the 1896 Dureya, as revealed to the readers of *The Autocar* magazine.

estimate of 500,000 spectators is particularly impressive considering the bad weather on the day. A banner hung across the road said 'Success to the motor car', and another at Brighton even paraphrased Napoleon (something that should have appealed to Harry Lawson), reading 'Centuries look down upon this thy immortal ride'. Some of the press certainly seemed to accept that the motor car was the vehicle of the future, even if 'not of the immediate present', as one of them put it. Others however, like the local Brighton papers, declared that 'all motor cars are vanity' and that 'as a public spectacle (the run) was a dreary failure', and, even a year later, some still doubted the importance of the car: 'The year that has elapsed…has not bought the motor so triumphantly into vogue as was at one time expected.' (*Illustrated London News*, December 1897)

Typically, Lawson declared the Emancipation Run a great success. Equally typically, he did so in a press release issued two days before the event itself. He also seized the moment to declare a third interim dividend of no less than 110 per cent for BMS (payable in Great Horseless Carriage Company shares, naturally) raising the year's dividend on this company with almost no visible product to 150 per cent. This was the prelude to him launching, within days of the run, two further promotions: the London Electrical Cab Company and a 'reorganisation' of BMS to enable it to buy yet more patents, despite it being already (as they claimed) the 'sole owner of all the principal motor car master patents of known and admitted value.' Together these two scams sought to relieve the British public of £1.15 million.

But the liberalisation of the laws on motoring did not bring about the forecast boom in car ownership. The leading continental manufacturers were already struggling to keep up with domestic demand, and the one domestic organisation that should have been able to deliver the new vehicles needed – Lawson's – was demonstrably incapable of doing so. As Lawson played for time the media went for the jugular. At a time when Lawson should have been cashing in on the new legislation by producing cars, shares in Lawson-owned companies began to collapse. As Nicholson put it:

> By the end of 1897 the empire built up by Lawson and his associates had fallen apart. Of all the companies with which they had been linked, only one or two offered even a trace of comfort to their shareholders.
>
> (Nicholson, page 470)

Many of the great and good who had rallied for Lawson when he appeared to be the only act in town promoting motorised travel slipped from his clutches. The Motor Car Club was denuded of membership as many went over to the genuinely independent Automobile Club of Great Britain and Ireland (later to become the RAC, modelled on the Automobile Club de France). Lawson's patent rights passed to others and were finally undermined when the validity of one of his key patents was successfully challenged by the Automobile Mutual Protection Association, a body set up by car makers to take on the monopolists, much as Henry Ford would do in America. By 1904 Lawson had earned himself twelve months' hard labour in prison for conspiracy to defraud in relation to another of his less substantial enterprises.

Meanwhile motorists had more pressing concerns. Now that the right of the motor car to be on the road had been established, the motorist was able to argue that it was as much the duty of the horse-owner to prepare his animal for the age of the motor car as it was for the motor car to accommodate the horse. One car dealer wrote to *The Times* offering the following cure for nervous horses, which he even offered to administer free of charge to those living near his showroom:

> Give a horse a good smell of a motor car standing still, then set the motor going and make the horse walk round the car half-a-dozen times; afterwards, keep the horse still and run the motor car round him half-a-dozen times. One such treatment would generally be sufficient, but two or three applications would perfect his education.
>
> (*The Times*, 30 September 1899)

The
Vauxhall Ironworks Co., Ltd.

MOTOR & MARINE ENGINEERS.

12-14 H.P. CAR SPECIFICATION.

THE 3 CYL. VAUXHALL CAR.

This 1905 advertisement for Vauxhall dates from before their move from London to Luton, which became their home for so many years.

The 1896 legislation may have been an important step forward, but it was still seen as being far less liberal than the regime applied in parts of mainland Europe (for example, one correspondent reported to a newspaper that he thought a speed limit existed across France, but it was so universally ignored that he could not be sure). More or less from the day of the 1896 Act coming into force, lobbying began for a further increase to the speed limit. A correspondent to *The Times* makes the case:

> [in 1896] it would have been too much to expect that the legislature or the Local Government Board should have authorised a vehicle which had not hitherto exceeded the walking pace of a man with a red flag to traverse the roads at any speed within its powers before the eye of the foot passenger and the nerves of the draught horse had grown accustomed to the change. Today, however, the case is different. The motor car has become a common object of the Queen's highway, [and] peer and tradesman alike have recognised the value of the new means of conveyance…

> (*The Times*, 15 December 1900)

They were strongly supported in this by the then Lord Montagu, Member of Parliament for the New Forest. He campaigned for the radical cause of abolishing speed limits entirely, replacing them with the offence of 'furious driving' originally framed in an Act of 1835 to deal with reckless horsemen. He offered as a *quid pro quo* the introduction of number plates so that those guilty of furious driving could be more easily identified. Montague made the point in a 1903 article that:

> ...you can beat your wife, steal, get drunk, assault the police and indulge in many other crimes and felonies, and you will find it cheaper than to go at twelve-and-a-half miles per hour.
>
> (*The Times*, 2 November 1996)

This was true; speeding carried a fine of £10 at the time, compared with £5 for assaulting a police officer. The fine for speeding put it on a par with 'harbouring thieves knowingly' or 'selling poisoned grain'.

At the same time there were pressures in the opposite direction, to reduce speeds. Middlesex County Council in November 1900 proposed to the County Councils Association that the speed limit for cars should be reduced from 12 to 10mph, and further reduced to 6mph at curves or crossroads, or when nearing a horse, carriage or foot passenger. (The practicality of trying to apply this latter rule is mind-boggling – what in law would constitute *nearing*? And how would it be established in a court?) They further re-introduced the proposal that cars should be licensed and numbered.

Middlesex County Council got only a part of what they wanted. Montagu's Bill failed twice, due to lack of parliamentary time. On the third occasion the Government introduced a virtually identical bill of its own. But the Bill's proposal of an offence of furious driving was not adopted. Instead, the Motor Car Act of 1903 raised the speed limit for vehicles below three tons to 20mph (though again the Local Government Board was given powers to reduce this to 10mph where it was deemed necessary). But the Act also introduced the offence of driving recklessly or negligently. It further introduced the registration of vehicles (number plates started appearing from 1 January 1904) and of drivers – though at this stage there was not any kind of driving test. Prior to this one of the ways of identifying vehicles was by the pet names painted on them by their owners (Rudyard Kipling's Lanchester was called *Amelia*).

The registration of cars had been bitterly opposed by motoring interests, with improbable claims being made that it would cause motorists to give up the use of their vehicles entirely, rather than register. In an earlier letter to chief constables, who were leading proponents of the numbering system, the Automobile Club attempted to reduce the idea to ridicule:

> There is no doubt that a provision that all private vehicles, tradesmen's carts and bicycles, all persons who have at any time been convicted of an offence against the law, and all tramps should bear a conspicuous number before and behind, visible by day and illuminated by night, would greatly assist chief constables in the exercise of their duties.
>
> (*The Times*, 15 December 1900)

The introduction of licensing also gave us the first incidence of a personalised number plate. It is said that Earl Russell sat up all night to secure the registration 'A 1' for his Napier car. He had been a motorist since 1898 and would, in 1926, introduce a private Bill to make motor insurance compulsory for motor vehicles (opponents claimed motorists would be 'encouraged to recklessness' by being insured).

But, even after this change in the law, it only took until 1905 for the Government to set up a Royal Commission under Lord Selby to look at the working of the 1896 and 1903 Acts, taking into account practices across Europe. It heard evidence from those who claimed that motor cars made peoples' lives a misery; how (in the hands of doctors) they were life-savers, and every shade of opinion in between. Among the more influential witnesses were the Hon.

Despite a 10mph speed limit and very little other traffic on the road, this car (possibly a Napier?) still managed to crash in Guildford in January 1907.

Arthur Stanley MP, chairman of the Automobile Club of Great Britain and Northern Ireland, then regarded as the voice of the motoring elite, and William Rees Jeffreys, the secretary of the Motor Union (the organisation that spoke for motorists of 'other ranks'). There were complaints about the 20mph speed limit, and a claim that the police practice of setting speed traps outside towns and villages was encouraging bad motorists to speed within the built-up areas. There was also a lively debate about the controversial use of horns, in which it was alleged that failure to sound them could lead to a motorist being held liable in the event of an accident. In theory the provisions of the 1903 Act were only supposed to remain in force until 1906, but pressure of parliamentary business meant that it was not until 1919, when the Ministry of Transport was set up, that motoring law was comprehensively reviewed.

But it only took a general election for politicians to appreciate the real purpose and value of the motor car, when a Mr Harvey Du Cros made one of the few Unionist gains in the Liberal victory of 1905 by mobilising a fleet of eighty cars to carry his voters to the polling station.

One of the biggest themes to emerge from the review of the legislation was the need for well-surfaced and dustless roads. The case for road improvements had been argued well before the advent of the motor car. The Cyclists' Touring Club (founded in 1878, with a membership of 60,000 by the time of the Emancipation Run) had pursued the cause vigorously. But when the growing band of motorists began to add their voices, results began to come in. In his 1909 budget, Lloyd George set up a Central Road Board. When the Board opened for business in 1910 it was the first time since the Romans left Britain in around AD 410 that the Central Government had taken any responsibility for the state of the nation's roads. Its works were to be funded by a 3*d* a gallon tax on petrol and a graduated tax on motor vehicles, ranging from two guineas for cars of less than 6.5hp to forty guineas for those in excess of 60hp. All the proceeds from these taxes were supposed to go to road improvements.

By 1906 there were sufficient cars for Britain's roads to be described as 'crowded' on Bank holidays. An independent tenth anniversary review of the Locomotives on Highways Act was undertaken, and one of those giving evidence was a Cambridge academic, G. Lowes Dickenson:

> For some ten years the people of this country have been groaning under a public nuisance which increases by the day until it has reached a malignity and magnitude altogether unprecedented. Their property has been depreciated; their security invaded. The amenity of the country has been indefinitely impaired and the discomfort of the town indefinitely increased. The citizen who does not motor has become a kind of outlaw on his own highways.

But even by the time of the 1911 Census the balance was still strongly in favour of the horse. Nationally, the Census showed a total of 471,000 people in the occupation of vehicle driver. Of these, 404,000 were working with horses and 17,000 with trams and other public transport, leaving just 50,000 for motor vehicles.

THE ECONOMICS OF THE BRITISH CAR INDUSTRY

> The motor car should have been British…you first invented it in the 1830s, you have the mechanics and the roads.
>
> (Thomas Alva Edison, 1901)

> …in the year 1900 France exported automobiles to the value of £290,360; but in 1902, the corresponding total had risen to £1,062,040. It is also said that 180,000 men are engaged in the industry in France, the wages of whom reach, in the aggregate, close on £13 million per annum. It is further stated that we are purchasing motor cars from abroad at the rate of about £3,000 in value each day, or about a million pounds a year.
>
> (*The Times*, 30 January 1903)

The British were one of the last of the major European nations to develop a large-scale motor industry. Some reasons for this are obvious, the main one being the crippling regulations limiting the use of cars in Britain before 1896, which provided no incentive whatsoever for a British motor car industry before that time. Another was the relative unattractiveness of the motor industry, which was something of a lottery prior to a broad consensus being reached on the main principles of the construction of a motor car – though this did not seem to have so great an inhibiting effect on continental Europe.

A further reason must have been the corrosive effect Harry Lawson had on the public perception of the motor car industry. We have seen that he and his companies were incapable of supplying the demand for cars, so why didn't other more honest British companies spring up to fill the gap? A number tried. With some, the product simply was not good enough – 'grossly underpowered' and 'dirty running' was how Nicholson described the kerosene-powered cars of Petter, Hill & Boll and Roots & Venables. But there was another problem that manufacturers, both good and bad, faced. Lawson had done for them all, so far as finding investors was concerned: 'The honest inventor…finds the money market as bare for him as a field of sprouting wheat after a flight of locusts.' (*Engineering*, November 1896) Shares in motor companies had generally become 'almost unsaleable'. If anything, interest in the development of the British motor car industry waned for a while after the liberalisation of the law.

Hough suggests two further reasons why the trade in motor vehicles might have been so one-way after 1896. His first argument is that the British were more a people of commerce than of manufacture. This seems odd given the country's wealth of engineering skills – both

The 1896 Roots & Venables' Oil-Motor Carriage. *The Autocar* commented (with commendable understatement): 'the vehicle is not so attractive in appearance as could be wished, but doubtless this will be improved later.'

Gottlieb Daimler and Armand Peugeot came to England to acquire those skills. The second more plausible reason is that other countries had far more restrictive tariffs than Britain, making it easier for them to sell their goods here than vice versa. There is also the suggestion that the requirements of the European market were rather different to those of Britain – customers in, say, France and Germany required cars capable of far higher speeds than were legal in this country. Nonetheless, after a shaky start, a total of 221 British firms went into car making between 1901 and 1905 (somewhat undermining Hough's first argument), of which no less than 90 per cent had left the industry or ceased trading altogether by 1914 (which may in turn vindicate Hough). A general liquidity crisis in 1907 did not help the development of the industry. Up to that time the major British companies had a much safer bet in manufacturing things like armaments, and motoring attracted far more interest from company promoters and speculators like Harry Lawson than from actual manufacturers.

Despite being left at the starting gate, the British industry did a great deal of catching up before the First World War. Between 1909 and 1913 French car production increased by about a third, while British grew three-fold. In the latter year British production (including commercial vehicles) was 34,000, compared with 45,000 in France and 23,000 in Germany. But total European production was still less than a quarter of that of the United States. By this time Britain had one of the highest densities of car ownership in Europe, with one car for every 165 people, compared with France (1:318) and Germany (1:950). But our level of car ownership still paled into insignificance against that of the United States, where 1.26 million cars were in use (one for every seventy-seven Americans). Britain had by then become the largest European market for cars, and France the main beneficiary of this market, for the British still had much to learn about producing small, cheap cars in large numbers. To illustrate the balance of trade, in 1907 the British car industry built 11,700 cars, of which 2,441 were exported. By comparison, 5,530 foreign cars were imported – over a third of total UK sales.

Too much of the British car industry was still concentrating on the coach-built perfection end of the market. They were not taking advantage of the new machinery that was becoming available, or the new manufacturing techniques, such as interchangeable parts, assembly lines and division of labour, that Henry Ford was to use to such good effect at Detroit and at Britain's Trafford Park. But Church questions whether the French were really much better than us in this respect. He could find little evidence at this time of better performance, in terms of low-cost high-volume production, on the other side of the Channel. There, as in Britain, many of the companies also tried to cater for a relatively small and very stratified market by building a wide range of products in relatively small batches, rather than running a continuous assembly line. Peugeot (the largest volume French manufacturer in 1913, with 5,000 units) was not so far ahead of the largest British company (Wolseley, with 3,000). For comparison, Ford was by then producing over 200,000 units a year (7,310 being assembled in the United Kingdom). Nor does he think that the French had the advantage of superior engineering skills, since only about one in five of the leaders of the French motor industry had a high-level engineering qualification. Finally, he concludes that it was not better marketing skills that earned the French an advantage in the British market, since it was British agents who were largely responsible for promoting French cars here. In the absence of any obvious distinguishing features between the two, Church concludes that the French had simply benefited from the head start given to them when Daimler sold them the rights to manufacture his engines in 1888. From then it would be a further eight years before it became practicable even to use a motor car on British roads.

But Henry Ford had seen the potential of the British market, even if many of the home-grown competition had not. Percival Parry, who had for some time been his agent, importing Model T Fords into the United Kingdom, persuaded him to manufacture Model Ts in England from imported kits. The factory at Trafford Park, near Manchester, opened in 1911 and was soon turning out cars costing £135 for the runabout and £150 for the touring model. The British press rather snootily dismissed them as 'cheap and nasty'. Cheap they certainly were, but nasty? The Model T had a relatively large (2.9-litre) engine that produced quite a lot of torque (albeit with only 20bhp). This was a disadvantage in Britain where cars were taxed on a formula which discriminated against vehicles with a large piston area. But despite its crudity of design and unsophisticated engineering, its ruggedness, reliability, simplicity and excellent after-sales service won it many friends.

By 1913 Model Ts accounted for over 60 per cent of British sales of cars costing less than £200. In that same year their principal British rival appeared in the form of the Morris Oxford. Most of the components of the Model T were made by Ford themselves, but Morris's approach to car building was exactly the opposite: to buy in virtually every component from an outside manufacturer. Only as the business grew after the First World War did Morris begin to buy up suppliers, or establish new ones nearer his main Oxford factory. The basic Morris Oxford model cost some 25 per cent more than the comparable Model T, at £175. But its selling point was its higher standard of engineering and finish and a more efficient, higher-revving engine. From 1915, its smaller-engined (1.5-litre) successor, the 'Bullnose' Morris Cowley became a serious competitor to the Model T, finally overtaking it in UK sales in 1924. Unlike the Model T, the Morris was specifically designed to take maximum advantage of the British car taxation policy.

THE BATTLE OVER SPEEDING

To satisfy the requirements of road traffic, ease in stopping and starting, rapidity in changing speed, and handiness in reversing are among the most essential qualifications for a motor. It is not necessary that the speed should be high; an average of 10 to 12mph would meet all ordinary requirements, while on rising ground a slower speed would suffice.

(The Times editorial, 7 December 1896)

The cost of motor cycling has been stated as a third of a penny per mile. Of course, it depends on the nature of the road. The figure given may apply to an ordinary stretch of highway, but if the scenery of the mile in question happens to be interspersed with policemen the cost is greatly enhanced forthwith.

(The Autocar, 1899)

Speeding was not a problem that emerged with the motor car. The earliest recorded British speeding fine dates from 1826, when a man was arrested near Tonbridge in Kent for doing 8mph in a built-up area restricted to 2mph. It emerged as a major issue after the passing of the 1896 Locomotives on Highways Act. Some authorities, even then, were quite open in seeing it as a valuable source of revenue. A single county bench took over £1,000 in motoring fines in a single week. Furthermore, the battle to re-introduce the red flag had by no means been abandoned after 1896. The vested interests involved in such a move should not be underestimated. In 1905 the value of horse-related industries to the national economy was put at £4,000 million.

As we have seen, the British maximum speed limit was further increased by the Motor Cars Act of 1903 to 20mph, but by then even quite modest cars were capable of 40mph or more, and many were caught speeding and subjected to substantial fines (for the day) of up to £15. In some cases it was apparently difficult for the cars to go slow enough – even in first gear – to keep within the urban speed limits laid down by some authorities. After he and his fellow motorists had been subject to a whole string of fines by the Paris police (some of whom apparently were trying to enforce those speed limits, notwithstanding the liberalism reported earlier), Arthur Rothschild wrote to Daimler asking if their cars could be equipped with a device that would slow the engine down to 4 or 500rpm. This would enable the car to keep within the French 8 and 10kph limits.

In the days before speed camera technology the method of detection consisted of a measured half-mile and a policeman concealed in a ditch at the end of it, with a telescope and a stop-watch. Electronic means of timing were tried early on, but were found to be so hopelessly erratic that they could not be relied upon for evidence. In some cases the prosecution was based on no more than a police officer's unsupported estimate of the defendant's speed. Motorists soon mobilised against what was seen as an infringement of their liberties. Expensive lawyers were bought in by affluent drivers to make a mockery of the country police constables trying to secure a conviction. Even where they succeeded in doing so, those courts with an anti-motoring bias were still prepared to convict on insubstantial or discredited evidence. The following cases were reported in *The Times*:

In August 1900 a Miss Vera Butler was charged with exceeding the speed limit. She claimed she was only doing 7mph, but the constable giving evidence estimated her speed (visually, without the benefit of measuring equipment) at 14mph, just 2mph above the legal limit. On the basis of this flimsy evidence she was fined £2 plus costs.

In September 1900 Lord Carnarvon was sighted by a policeman in plain clothes leaving his home at Highclere Castle in his car, travelling at what the officer said was an excessive speed. The officer pursued him on his bicycle at the legal limit of 12mph (we are not told how he knew his own speed), but was left well behind. Another officer later timed him over a stretch of road which, when measured, led them to calculate that he had been travelling at 24mph, raising a huge cloud of dust and skidding as he rounded corners. The officer said he had received complaints from local people about the speed at which Lord Carnarvon drove his car. Despite one of his Lordship's servants giving evidence that he had been in the car at the time, and that it had not exceeded 12mph, the defendant was fined ten guineas plus costs.

Entrapment was not unknown. A case was bought before magistrates at Slough in 1899 of a motorist arrested for failing to stop when signalled to do so by the driver of a horse and cart. When it transpired that the carter, and the accompanying policeman, had been waiting there for forty-five minutes to entrap somebody, the case was dismissed, but the motorist was not awarded costs, nor was any criticism made of the police practice.

It was not surprising that motorists formed organisations to protect their interests. As we saw, in keeping with the rather elite membership of the early motoring fraternity, the Automobile Club (founded in 1897) was very grand. Its president was the Duke of Sutherland and its membership included Lord Montagu, members of the Royal Family and two Prime Ministers – Messrs. Rosebery and Balfour. Members had to pay an additional fee of 15s for the registration of their chauffeur, and the organisation operated as a gentlemen's club (first from Whitehall Court, then later from Piccadilly). Being a gentlemen's club, women were naturally not admitted, so they formed their own Ladies' Automobile Club, based at Claridges. The Automobile Club – which subsequently became the Royal Automobile Club (RAC) – also formed the Motor Vehicle Users' Defence Association, to protect its members from what they saw as the persecution of the law.

The Motorists' Union (the body founded by Charles Jarrott, which later became the Automobile Association) became an important part of the motorists' retaliation, bringing a more militant approach to their opposition. They produced maps to show where the favourite speed trap locations were, and by 1908 these were being regularly published in *The Autocar*. They would also send volunteer cycle patrols down key routes (London to Brighton was first, followed by the Portsmouth Road) to see where the police were in action on any given day, and then go a short distance up the road to forewarn their motorist members by waving a flag as they approached. By 1906 some 100 patrols were active at peak motoring times. The AA badge for cars was adopted, so that motorists who were not members could be declined this advantage.

One of these scouts even testified in court that he had been cycling uphill alongside the car, which therefore could not have been going as fast as the police officer claimed. Not only was his evidence not believed, but he was separately prosecuted for perjury, the charge only being dismissed on appeal. In a well-publicised test case in court, it was held that these 'spies' were not obstructing the police in their duties, so the authorities tried changing their methods with a view to bringing these patrolmen within the scope of prosecution. Given the possibility that they might be prosecuted for their actions, the patrolmen adopted the tactic of inaction. It was announced that they would not provide their habitual salute to their members if there was a speed trap ahead. Even the most ingenious of courts could not find a way of prosecuting them for failing to salute a club member. For those members who were not forewarned, the AA introduced free legal representation at court for motoring offences.

But the AA was more than just an anti-speed limit pressure group. They took a number of initiatives to ease the lot of early motorist. They installed some of the earliest traffic warning signs, following a tradition started by the Cyclists' Touring Club ('Dangerous Hill' apparently being one of the Cyclists' Touring Club's favourites). They brought in branded village name signs at a time when newspaper campaigns were being conducted to get village post offices to paint their village name on their shop signs as an aid to lost travellers. They developed the first real cross-Channel car ferries in 1908 – prior to that, most cross-Channel traffic went by rail and foot. And they provided sentry boxes for their scouts, which later evolved into the AA boxes equipped with a telephone available for use by members in the days before mobile telecommunications.

Just as the railways had before them, so the motor car had its dramatic impact on patterns of settlement. Where the railway age had promoted the building of settlements within easy reach of stations, so the personal mobility of the motor car opened up a new world of possibilities. Batemans, the house that Rudyard Kipling bought in Sussex, was well beyond the convenient reach of a station by horse-powered transport. But, thanks to the Lanchester he drove, he was able to acquire the property at a knockdown price before the vendor realised that a dramatic shift had taken place in the boundaries of the property market.

TAXATION

One thing the governments of the day will have welcomed about the motor car is the opportunity of opening up new streams of tax revenue. Herbert Asquith said that a 'tax on motor cars would be almost an ideal tax because it is a luxury which is apt to degenerate into a nuisance.'

The 1896 Act required the owners of 'light locomotives' weighing less than one ton to pay £2 2s for a licence, and £4 4s for those over a ton. Then the RAC came up with a method for classifying cars in competitions, which in 1906 was adopted by the Society of Motor Manufacturers and Traders for cataloguing them. It was then adopted by the Treasury in the Finance Act of 1910 as the basis for their taxation. It also later became the basis for calculating the initial cost of the car and insurance charges. The formula was based on the number of cylinders and their bore. It took no account of the stroke of the engine (and hence its capacity), nor of its compression or the speed at which the engine ran (and hence the power it could produce). As compressions and running speeds increased, it became an increasingly misleading way of classifying engines. It encouraged British manufacturers to produce small-bore, long-stroke, high-revving engines in an effort to minimise the tax and other cost burdens for their customers. This was applied into the 1930s, and did much harm to the British export industry since these were not especially efficient engines, nor were they what the export markets wanted.

CHAPTER 9

THE AMERICAN
EXPERIENCE

These machines (motor cars) are much in vogue among the English aristocracy, and will undoubtedly soon be here.

(*Detroit Free Press*, 1896)

There is something uncanny about these new-fangled vehicles. They are all unutterably ugly and never a one of them has been provided with a good or even an endurable name. The French, who are usually orthodox in their etymology, if nothing else, have evolved 'automobile', which being half-Greek and half-Latin is so near to indecent that we print it with hesitation.

(*New York Times* editorial, 1899)

The motor car was as slow to emerge in America as it was in England, and its early development took a rather different path to mainland Europe. What is normally recognised to be the first American car appeared in Springfield, Massachusetts, in 1893. It was built by two cycle mechanics, Charles and Frank Dureya, who took their inspiration from a description of Benz's first car in the *Scientific American* magazine. In 1895 there were just 300 motor cars in the whole of America, and, by the following year, the Dureya brothers were the country's largest home-grown manufacturers, with an output of just thirteen cars, some of which, as we saw, actually took part in the London to Brighton Emancipation Run. Others by this stage, including Henry Ford, had produced prototype cars. But these tended to be more experimental vehicles than commercial products. The first serious manufacture of cars did not take place in the United States until 1897, when Pope, Stanley and Winton all started producing cars for sale. Ransom E. Olds also set up his Olds Motor Works in that year, but it would be another two years before he was in a position to begin seriously manufacturing his Oldsmobiles.

The first official record of American car production numbers dates from 1899, when around 2,500 vehicles were built. Most of them came from one of ten firms based in New England, and most were powered by steam or electricity. From there the volume of manufacture, and the

The 1893/4 Dureya, possibly America's earliest car.

number of manufacturers, grew swiftly. By the turn of the century, when America's population of cars numbered some 8,000, there were seventy-two manufacturers. As markets and production gravitated towards the mid-west, manufacturers concentrated particularly in south Michigan. By 1903/4, when the Ford Motor Company was established as a manufacturing concern, there were twenty-two manufacturers based in this region producing between them 42.1 per cent of the national output of cars (21,675 in that year).

Car ownership rapidly became, if not the norm, relatively commonplace among American families. In 1902, according to one estimate, the United States had one car per 1.5 million citizens. By 1905 this was reduced to 1:65,000 and by 1908 to 1:800. By 1908, when Britain was turning out 11,700 vehicles, the USA produced 65,000. Production passed a quarter-of-a-million in 1911, half-a-million in 1913, and by 1915 (when most of the European car industry was preoccupied with other matters) a million cars were being produced in America. By 1912 the city of Detroit alone was building some 82,400 cars a year – more than three times the entire output of the British industry.

The Oldsmobile factory proved to be something of a finishing school for the future leaders of the American car industry. Engines for the first Oldsmobile cars were designed by Henry Leland, who later went on to head both the Cadillac and Lincoln car companies, whilst the early Oldsmobile transmissions were produced by the Dodge brothers (who also worked with Ford). Another car manufacturer to cut their teeth at Oldsmobile was Maxwell and Hupp,

The unsuccessful Sears Roebuck model range. By the time this advertisement appeared in 1912 they were years behind the competition.

SEARS MOTOR CARS

——Better Than Ever——

$335⁰⁰ TO $535⁰⁰

No. 21H333
Model "H" Price $385.00

Ten-Day Trial on All Sears Motor Cars—Satisfaction Guaranteed

Most liberal selling proposition known in the automobile business. Use the car for ten days and satisfy yourself. If it is not as we represent it, return it and we will give you back your money, including all freight charges.

SEARS CARS have made good in the hands of thousands of owners. We sell direct and save you from $100.00 to $150.00.

ECONOMICAL

PRACTICAL

RELIABLE

No. 21H343
Model "P" Price $495.00

No. 21H777
Model "L" Price $495.00

SEND FOR OUR 1912 AUTOMOBILE ——CATALOG——

It will pay you to investigate the Sears before purchasing an automobile. We have issued a HANDSOME SPECIAL AUTOMOBILE CATALOG describing our complete line, and we will be glad to send this free upon request to anybody writing for it, together with our Testimonial Book, "What Sears Owners Say." Write for them today.

whose Hupmobiles were a significant player in the industry either side of the First World War. 1908 was a significant date in the development of the American car industry, when Buick and Oldsmobile formed the original core of what was to become the giant General Motors. They were joined by Cadillac and the less well-known Oakland (later to become Pontiac) the following year.

Many other companies, large and small, entered the business, most of whom have since been forgotten. The American public visiting the 1904 New York Automobile Show were warned against the proliferation of small-scale manufacturers:

> …a word of warning should be spoken against a certain type of exhibitor whose sole plant, capital, output and experience is represented by the one, solitary machine he had on exhibition, but who nevertheless does not hesitate to solicit orders, in the hopes that he may place enough of them on his books to guarantee the purchase of a few more tools and the employment of a few more hands at his so-called establishment.
>
> (*Scientific American*, 30 January 1904)

One significant failure whose company is most definitely not forgotten today is the Sears, Roebuck Company, who launched the Sears Auto Buggy in their 1908 catalogue. They hopelessly misjudged their market, producing a lightweight, flimsy and underpowered vehicle that was already ten years out of date and was entirely unsuited to the rough roads most American drivers had to contend with. Despite the fact that it undercut the price of the Ford Model T, introduced the same year, they had lost some $80,000 by the time they stopped manufacturing it in 1912. Thereafter they made money rather more successfully from the motor industry by supplying spare parts to other manufacturers.

ALTERNATIVE POWER

America favoured steam and electric cars over the internal combustion engine for much longer than their European counterparts. American new vehicle registrations for 1900 were made up of 40 per cent steam cars, 38 per cent electric and only 22 per cent petrol. Each had its advantages and disadvantages:

STEAM CARS: We looked at the development of the steam road vehicle, principally in Europe, in earlier chapters. By the 1890s steam engines were powerful, smooth running, fast and relatively clean. As we have seen, one of their major disadvantages – the time it took to heat up the water – was largely overcome by the invention of the flash boiler by Leon Serpollet in 1889, which made it little slower in its getaway than its petrol-based competition. Its other disadvantages – cost, weight, the tendency to freeze up in cold weather and the need for frequent stops to take on water – were not so easy to overcome. Nor was the fear of an explosion.

ELECTRIC CARS: Electric cars were a very different package. Electricity was instant, controllable and clean to use. It was easy to recharge from a domestic electrical supply, or at one of the

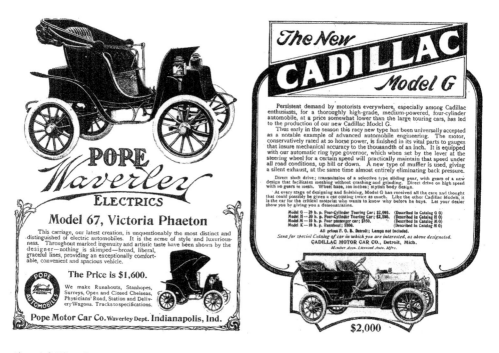

Above left: Electric cars were initially more popular in America than the petrol variety, but by the time this advertisement for Pope appeared (1907) they had passed the peak of their popularity. *Above right:* The 1907 Cadillac offers lights as an optional extra.

'charge garages' that were opening in many of the towns and cities. One drawback was that a full recharge took some three hours but, so long as the only decent roads in America were largely confined to urban areas, and inter-city travel by car was not a realistic prospect, at least the electric car's relative lack of range and low speed were less of an issue. This was particularly the case when we consider that most cities at the turn of the century were subject to an 8mph speed limit.

Electric cars also had the great advantage that they were the easiest of all to start, which appealed to many female drivers – and, no doubt, a number of males – until the introduction of the electric self-starter for petrol engines.

Electric cars have a long history which pre-dates the working internal combustion engines of Daimler and Benz. The first primitive electric carriage was demonstrated by a Scot, Robert Anderson, at some time between 1832 and 1839, while a Dutchman, Sibrandus Stratingh, designed a small-scale electric car in 1835. An electric bicycle, designed by an Austrian, Franz Kravogl, was displayed at the Paris World Exhibition of 1867. Improvements to the storage battery in 1865 and 1881 made electric vehicles a more practicable proposition, and a Frenchman, Gustave Trouve, displayed a working three-wheeled car at the International Exhibition of Electricity in Paris in 1881.

At first electric road cars could match the performance of at least some of their petrol-powered counterparts. An electric car with a top speed of 20mph was no slower than, say, a petrol-driven Oldsmobile runabout, and was cleaner and easier to start. Some specialised electric cars even set world speed records. Perhaps the most famous of these was *La Jamis Contente*, whose driver, Camille Jenatzy, was, as we saw, the first man to break the 100kph (60mph) barrier in 1899.

But several factors eventually told against the electric car. The technology of electric cars did not develop as quickly as internal combustion. They were soon out-performed. Their advantage in ease of starting was virtually lost with the development of the electric starter by Cadillac, and the mass-production of the Ford Model T made internal combustion highly competitive in price. The cost of electric motoring was also increased by the wholesale introduction of an alternating, rather then direct, current, making it necessary for electric cars to have a costly converter.

Also, as a network of decent inter-urban roads grew in the years leading up to the First World War, the limitations of the electric car were gradually thrown into even sharper relief. These had been apparent much earlier in continental Europe where, as *Scientific American* reported in 1895, the Count de Chasseloup-Laubat had to hire a train 'at considerable expense' to place fresh supplies of accumulators at different stages along the route between Paris and Bordeaux to enable him to take part in the inter-city race of that year in his electric car. In America the electric car reached the pinnacle of its popularity in about 1905, and gradually declined thereafter.

One important contribution America made to the development of the motor car was in terms of its mass production. One obvious reason for this was the existence of a large and relatively undifferentiated market. Brandon suggests that there was also a historical context for mass production developing in the United States. After the American War of Independence the British imposed a ban on exports of machinery to the United States, hoping to demonstrate America's economic dependency on the mother country. Instead, the Americans developed their own technologies, which were in any case better suited to their circumstances. In the absence of skilled labour they resorted instead to the sub-division of tasks into simpler elements, which in turn lent themselves to mechanisation. The idea of the standardisation of parts and mass production was applied to the manufacture of armaments during the American Civil War (the founder of Cadillac and father of interchangeable car parts, Henry Leland, worked for gun-maker Samuel Colt before going into car manufacture). But this process had its origins even earlier in France, having been developed there for the manufacture of muskets by an engineer named Le Blanc and brought back to America by Thomas Jefferson, one-time American Ambassador to France.

Above left: For many car manufacturers a transcontinental trip across the United States was the ultimate test of their products. *Above right:* Studebaker hedged their bets with an electric car for local journeys and a petrol model for touring.

Henry Ford was certainly not the first to apply these principles to the manufacture of the motor car. Oldsmobile and Packard both did so to varying degrees before him. But nobody honed it to such scientific perfection as Ford. However, Ford did not initially seek to serve the mass market. The cars he made between 1903 and 1908 cost an average of $1,600, whereas two-thirds of the American cars built in 1903 cost less than $1,375. Other manufacturers, notably Ransom E. Olds with his Oldsmobiles, were already catering for the bottom end of the market. However, many of them sacrificed durability for cost. It wasn't until 1909 that Henry Ford had what Brandon refers to as his epiphany. He described his new mission in his ghost-written 1925 autobiography, *My Life and Work:*

> I will build a motor car for the great multitude. It will be large enough for the family but small enough for the individual to run and care for…But it will be so low in price that no man making a good salary will be unable to own one – and enjoy with his family the blessings of hours of pleasure in God's great open spaces.

Ford's first inexpensive car was the Model N, which he initially brought out at $500. But without the economies of mass production he was unable to sustain that price. It was with the launch of the Model T that sales really took off. The level of sales they reached were such that some brokerage firms refused to buy into municipal bonds issued by those cities which had, in their view, 'too many automobiles in proportion to population'. The grounds for this appeared to be that they thought too many families were over-stretching themselves to buy motor cars, and that their resultant inability to pay their municipal dues would make those authorities bad credit risks.

SOME AMERICAN PIONEERS

It has been estimated that between 2,000 and 3,000 American companies have set up as motor car manufacturers – with varying degrees of seriousness – over the years. Here are the stories of just a few of the most important contenders in the years before 1914.

BUICK: Buicks have been described as the 'upper middle-class ground of high quality yet unostentatious cars'. David Buick was a Scottish-born immigrant, bought to America as a baby, who made money in plumbing before turning to motor manufacture. (Two of his minor claims to fame were as the inventor of the white porcelain bath and the lawn sprinkler.) Like Henry Ford, he developed such an interest in the internal combustion engine that it distracted him from making a living, and his plumbing business folded. Like many other inventors (but in this respect unlike Henry Ford), he was described as 'a complete flop' as a businessman.

He set up the Buick Auto-Vim and Power Company in 1899 with the aim of supplying internal combustion engines for agricultural applications, but again spent all his time experimenting with cars. He ran out of money early in 1902 with only one prototype car to show for his labours. Undeterred, he founded the Buick Manufacturing Company, to manufacture cars and supply engines to other car makers. Once again he had run out of money by the end of the first year. But in that time he had developed the revolutionary overhead-valve engine on which almost all modern engines are based.

With the help of a loan from Benjamin Briscoe who must, with Buick's track record, have been a very trusting friend, the Buick Motor Company was formed in 1903. This company was said to have produced sixteen cars in its first year, though proof of this is elusive. We do know that thirty-seven Buicks were sold the following year. But the company was taken over that same year by James H. Whiting, and a wealthy carriage manufacturer, William Durant, joined the company in a managerial role. By now Buick had become something of a menace, constantly trying to halt the production line to tinker with the design. But despite his best efforts, by 1905 the company was worth $1.5 million. Buick was bought out for $100,000, most of which he promptly lost in oil speculation. He died in relative poverty in 1929, never having been able to afford one of the cars which bore his name.

Durant's business skills soon made Buick the largest car manufacturer in America, and from there he began a series of acquisitions that were to become the giant General Motors Corporation. Within the strict pecking order of prestige that existed in the corporation, Buick eventually became second only to Cadillac as a luxury model. Like Ford's Model T (which was not entirely correctly famed for coming in 'any colour so long as it is black'), Buick had its own trademark colour, an off-white known as Buick grey, in which many of its cars were finished.

CADILLAC: When Henry Ford broke up the partnership that was the early Henry Ford Company in 1902, Ford's partners brought in an engineer, Henry Martyn Leland, to advise them on the sale of the factory's assets. Leland's background had been in making precision gears, steam engines and marine petrol engines. More recently he had been under contract to Oldsmobile producing engines for their curved dash model. He had developed an improved version of their engine that gave 23 per cent more horsepower, but it was rejected by Olds, partly because of the time it would take to re-tool to build it.

Leland's advice to the Ford backers should, therefore, not have been entirely unexpected. He told them to continue the business as a going concern. He had his reliable single-cylinder engine, which was fitted into a virtually unchanged Ford Model A, and the first production model emerged from the factory in October 1902. It was shown at the New York Auto Show in January 1903 and won 2,286 orders before being declared 'sold out' mid-show. The new marque was named after the French explorer who founded Detroit in 1701, Antoine Laumet de la Mothe, Sieur de Cadillac. Fortunately for future marketing prospects, the name was shortened, or the badge would have taken up the entire front of the car.

From the start Cadillacs were set apart by the quality of their design and engineering. Their well-publicised interchangeability of parts, at a time when many rivals' components required lengthy hand-finishing, is described elsewhere in the book. At the same time the earliest Cadillacs were very affordable, coming in at between $750 and $850. (But, having quoted these prices, it may be worth noting that some items were only available as optional extras at the time. These included a horn, lights and weather protection. How optional were they, even in those days?)

Their growth was temporarily curtailed when the Cadillac factory burned down in the spring of 1904. But whilst the mechanical development of Cadillacs progressed apace, the assembly of the bodies and chassis was pitifully old-fashioned and, in October 1905, a restructured Cadillac Motor Car Company was formed, to bring the progress of the two into line.

By about 1906 they recognised that the market for even the best single-cylinder cars was eroding, and they began their migration to the luxury end of the market. They started by focussing on high-volume single-line production to produce a high-quality, but still relatively inexpensive, model, the 'Thirty', costing $1,400. This sold almost 6,000 units in 1909, the year that Cadillac was acquired by General Motors.

Cadillac was responsible for a number of technological breakthroughs in car design. In the period in which we are interested these include a decompression device to aid cranking (1904), the first advanced electrical system combining cranking, lighting and ignition, using a powerful battery-driven dynamo (1912), the electric starter as standard equipment (1912) and the first large-volume production V8 engine (1914). The company also introduced the first fully enclosed cab as standard equipment (1910). In those days, across much of America, travel conditions precluded winter motoring. Orders for new cars were taken at winter car shows for delivery during spring and summer. Cadillac changed this; their year coincided with the calendar year, and their stock tended to be pre-sold. But they took this even further, so that by the summer of 1910 they had already announced their 1911 range, and had delivered a significant proportion of their output by the start of calendar year 1911.

CHEVROLET: Louis Chevrolet was a Swiss-born immigrant to America who had worked in the French car industry before emigrating – initially to Montreal – in 1900. He worked as a chauffeur to David Buick and as a racing driver, and was hired by William Durant to drive his Buick cars in promotional races. Durant went on to found General Motors, but in 1910 was

J.B. Bartholomew was a manufacturer of peanut and coffee roasters from Peoria, Illinois, who turned his hand to car manufacturing. The early prototypes (of which this is one) bore his name, but the production models from 1904 were branded as Glides – 'Ride in a Glide. Then decide'.

forced out of the company. He set about rebuilding his place in the motor industry through Chevrolet's designs, and a company to promote them was set up in 1911. By 1916 it was profitable enough for Durant to buy back a majority shareholding in General Motors. Louis Chevrolet, whose name is thought to come from the French *chevre a lait*, meaning a milk goat, giving some hint to the trade of his forebears (the Milk Goat Corvette will never seem the same again!) quit the company in 1913, apparently unhappy with Durant for taking his brand name down-market.

The origin of the company's stylised cross logo is also something of a mystery. It is apparently not a version of the cross on the Swiss flag of Chevrolet's birthplace. It is either a stylised version of the shape of Switzerland itself or, more mundanely, a copy of a motif Durant saw in the wallpaper of a French hotel.

DUREYA: The Dureya was the first American-built car with an internal combustion engine, though there appears to be some debate as to whether Charles Dureya (1861-1939) or his brother Frank (1870-1967), or both, can claim the credit for it. The first Dureya engine was built as long ago as 1886, but did not find its way into a car until 1893. It first ran on the streets of Springfield, Massachusetts, on 20 September of that year, and the brothers' first patent dated from 1895. As we saw, an output of just thirteen vehicles in 1896 was enough to make them America's largest volume car manufacturer for that year.

The company, however, had a very short life, as the two brothers fell out and the company was dissolved in 1898. Frank went to work with the Stevens Arms and Tool Company, eventually producing the Stevens-Dureya car, which continued in production until 1927. Charles moved to Pennsylvania where he produced a string of unsuccessful car designs. Attempts to introduce the Dureya brand to England also proved unsuccessful.

HUPMOBILE: Robert Craig Hupp, who had worked for Oldsmobile, Ford and Regal cars, began producing cars in his own right from 1909. The first model was a lightweight two-seater. Production grew rapidly from 1,608 in 1909 to 12,543 in 1913, but Hupp consequently fell out with his financial backers and sold his interest in the company. The company nonetheless continued to grow after the First World War, until losing its way and folding at the end of the 1930s.

LOCOMOBILE: The Locomobile Company was formed in 1899 from a takeover of the Stanley Steam Car Company. The name is a combination of 'locomotive' and 'automobile'. Their first products were, not surprisingly, steam cars. They were described as being unreliable, finicky to operate, prone to catching fire, slow to raise steam and short of range (needing to take on water every twenty miles), or, as Rudyard Kipling rather more memorably described them, 'a nickel-plated fraud'. Despite all this they were relatively cheap and proved very popular, selling over 4,000 between 1899 and 1902 alone. They became the first cars to have a military application when the British used them in the Boer War. Notwithstanding their intended role of towing searchlights and generators, and providing a catering vehicle, British troops found their boilers particularly useful for brewing tea.

By 1903 the company had decided that the future lay with the internal combustion engine, and they sold their steam vehicle rights back to the Stanley brothers. They became known for well-built and fast, but very conservative, luxury cars, and they sought to enhance their reputation by entering racing. Their 16.9-litre monster failed in the 1905 Gordon Bennett Cup but came third in the Vanderbilt Cup of the same year. In 1908 they became the first United States-built car to win an international race, taking first and third in that year's Vanderbilt. That was just about the full extent of their motor racing success.

OLDSMOBILE: Until it closed in 2004, Oldsmobile was the world's oldest surviving car manufacturer after Daimler and Peugeot. In their 107 years they produced some 35.2 million

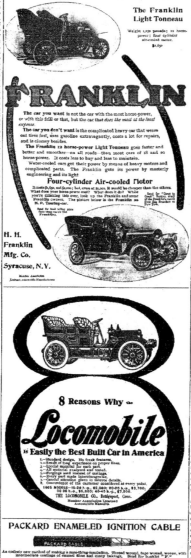

The Franklin Light Tonneau

Weight 1,650 pounds; 12 horse-power; four-cylinder air-cooled motor. $1,650

FRANKLIN

The car you want is not the car with the most horse-power, or with this frill or that, but the car that *does the most at the least expense*.

The car you don't want is the complicated heavy car that wears out tires fast, uses gasoline extravagantly, costs a lot for repairs, and is clumsy besides.

The Franklin 12 horse-power Light Tonneau goes faster and better and smoother—on all roads—than most cars of 18 and 20 horse-power. It costs less to buy and less to maintain.

Water-cooled cars get their power by means of heavy motors and complicated parts. The Franklin gets its power by masterly engineering and its light

Four-cylinder Air-cooled Motor

It costs $1,650, not $2,000; but, even at $2,000, it would be cheaper than the others. What does your horse-power cost? What does it do? While you're thinking this over, look up the Franklin and some Franklin owners. The picture below is the Franklin as H. P. Touring-car.

H. H. Franklin Mfg. Co. Syracuse, N.Y.

8 Reasons Why the Locomobile is Easily the Best Built Car in America

1.—Standard design. No freak features.
2.—Result of long experience on proper lines.
3.—Special material for each part.
4.—All material analyzed and tested.
5.—Forgings used instead of castings.
6.—Every part made interchangeable.
7.—Careful attention given to minute details.
8.—Convenience of the customer considered at every point.

1905 MODELS—15-20 h-p., $2,800; 20-25 h-p., $3,700; 30-35 h-p., $5,000; 40-45 h-p., $7,500.

THE LOCOMOBILE CO., Bridgeport, Conn.
Member Association Licensed Automobile Manufac.

PACKARD ENAMELED IGNITION CABLE

PACKARD CABLE

An entirely new method of making a sure-thing-insulation. Thread wound, tape wound, woven, with intermediate coatings of enamel time and time taking. Send for Booklet "F."

THE PACKARD ELECTRIC COMPANY, Warren, Ohio

The Locomobile – either 'the best built car in America', or, as Rudyard Kipling put it, 'a nickel-plated fraud' with a tendency to catch fire.

cars. Ransom E. Olds was a young man with engineering skills, who developed an interest in the automobile. He built a petrol-fuelled steam car as early as 1887. By 1896 he was driving one of his own petrol-engined cars in Lansing, Michigan, where it attracted interest – and investment – from financiers in Detroit. The Olds Motor Vehicle Company was established that year with a share capital of $5,000. Various models were under development when a fire destroyed most of the factory in 1899.

All that was saved was the prototype for a low-cost car – hence, their first model was a light American runabout, known as the 'curved dash', with an engine that turned over at just 500rpm or, as one wag put it, 'one chug per telegraph pole'. An interesting feature of the model was a single long spring on either side that extended from the front to rear wheels and doubled as part of the chassis. Not only did it work very efficiently, it was also cheap and easy to manufacture, helping to make the car very competitively priced at $650. This was,

The Oldsmobile Curved Dash was probably the world's first mass-produced car (though Olds left the invention of the moving production line to Ford). They sold from 1901 at $650 (though this one has the optional extras of mudguards and lights). In 1903 L.L. Whitman and Eugene Hammond drove one (at a top speed of 20mph) across the roadless wastes of America, from San Francisco to New York, in seventy-nine days.

however, the basic price, and, like the earliest Cadillacs, so-called extras included such essentials as mudguards ($10 extra), rear seats ($25), a hood and lighting. It nonetheless became the first high-volume mass-produced motor car. In 1901, the first year of its manufacture, 425 were sold, rising to 2,500 the following year and 6,500 in 1905. The curved dash also demonstrated its durability by completing a seventy-three-day journey across the United States from San Francisco to New York in 1903.

During the six years of its manufacture, Oldsmobile sold over 19,000 of their runabouts, and between 1903 and 1905 they were the world's highest volume car makers. They are credited with developing the first automobile assembly line in 1901. The one refinement it lacked, which Ford was later to introduce, was the conveyor belt. The cars were also popular in England, and were manufactured under licence in Germany. Just as the Model T Ford had its own music hall anthem, *Get out and get under*, so the Oldsmobile brand entered the American musical imagination through the waltz *In my merry Oldsmobile*.

Olds bought in many of his components for the runabout from outside suppliers. Supposedly identical engines were supplied by the Dodge brothers and Henry Leland (in his pre-Cadillac days) but, as we have seen, it is notable that Leland's superior engineering made his motors measurably more powerful than the Dodge variety. Olds, however, was more interested in turning the product out than in the finer points of engineering, and this led him to shy away from Leland's proposals for developing an improved version of the engine.

Unlike Chevrolet, who quit the company in protest at it being taken down-market, Olds resigned when the majority shareholders decided to go up-market. This move culminated in the 11.6-litre monster known as the Limited, in 1910, which cost $6,400, more than a three-bedroom house of the day. They sold just 725 of the vehicles in three years. For his part, Ransom E. Olds went on to found the REO Motor Car Company.

PACKARD: The Packard brand was born of a challenge. When the purchaser of one of Winton's early cars complained about it, the manufacturer testily suggested that, if the complainant thought he could do better then he should go into the business himself. That complainant turned out to be James Ward Packard, and he duly followed Winton's advice, rather successfully. (Enzo Ferrari later made a similar mistake with one of his customers, one Ferruccio Lamborghini.) By 1899 the Ohio Automobile Company, as it was originally known, had begun manufacturing, and from the start aimed at the top end of the market. At the time when an Oldsmobile runabout could be had for $650, Packard's range started at $2,600. Their cars attracted an elite following, both in America and overseas, with many heads of state buying them. It was from among their owners that they secured the funding to expand. A proud owner from one of Detroit's wealthiest families – Henry Bourne Joy – put together a consortium of backers that enabled the Ohio Automobile Company to transform into the Packard Motor Company in 1902. Their factory on East Grand Boulevard in Detroit was said to be the most modern car manufacturing facility in the world when it opened the following year.

PONTIAC: This started life as a company founded by Edward P. Murphy in 1907 under the brand name Oakland. They were built in the town of Pontiac, Michigan. Murphy sold a half share to General Motors in January 1909 and died in the summer of that same year, at which time GM bought the other half. The company found a place in General Motors' pecking order, below Oldsmobile and Buick, but above Chevrolet. The brand name Pontiac did not appear until 1926 and it was a further five years before the Oakland marque was discontinued.

RAMBLER: Between 1900 and 1914 this was the brand name of the Thomas B. Jeffery Company. Jeffery's cars were also known as 'Kenosha Cadillacs', after the town in Wisconsin where they were manufactured, in a former bicycle factory. The Rambler name first appeared on an 1897 prototype which got good reviews at the 1899 Chicago International Exhibition. By 1902 they were producing some 1,500 cars a year, making them the largest American car builder after Oldsmobile. They experimented with a steering wheel on one of their early models, but decided the technological competence required might be a little advanced for their customers, and so reverted to a tiller. Rather more positively, they were among the first manufacturers to equip their cars with a spare wheel and tyre. In the days when punctures were all-too-commonplace, this made roadside repairs much swifter, by doing away with the need to detect the puncture and patch the inner tube at the roadside.

On Thomas Jeffery's death, his son Charles renamed the marque Jeffery, in his father's honour. The company was bought by Nash in 1917, and it wasn't until the 1950s that the Rambler name was reintroduced.

WINTON: Alexander Winton was one of America's earliest car makers. He was a Scots emigrant who worked as a ship's engineer before going into manufacturing bicycles in 1891. He had already built two prototype cars when he set up the Winton Motor Carriage Company in 1897, based in Cleveland, Ohio. One of his earliest cars was recorded as achieving no less than 33.64mph around a Cleveland horse-racing track in May of that year, and he was able to add claims of reliability to those of speed by sending the car on an 800-mile test run from Cleveland to New York City. By 1899 his sales exceeded 100, making him America's largest petrol-engined car manufacturer of that year (though Columbia electric cars and Locomobile steamers still out-sold him). His early cars were very much of the horseless carriage type: the Phaeton of 1899 lacks only the front shafts of a horse-drawn buggy.

The company's reputation was boosted in 1901 when two members of the Vanderbilt family bought Wintons, but in the following year he lost his famous race with Henry Ford, described elsewhere in the book. However, a new model, the Winton Bullet, briefly and unofficially set a new land speed record of 70mph, and in 1903 one of his cars had the severest possible test of endurance. It was in a Winton that Dr Horatio Nelson Jackson made the first successful west to east transit of the United States, taking sixty-four days (including waits for spare parts to be shipped out by rail). The Winton Motor Carriage Company ceased production in 1924.

AND THE BIGGEST PIONEER OF THEM ALL, HENRY FORD (1863-1947)

In 1847 twenty-one-year-old William Ford, an Irish refugee from the potato famine, arrived in America aboard one of the immigrant 'coffin ships' to make a new life for himself. He made his way to Dearborn, Michigan, just eight miles from the growing city of Detroit where relatives had settled some years before. When William carved a farmstead out of the wilderness he could hardly have guessed that his son would become one of the richest men in America and a national icon.

That son, Henry Ford, was born during the American Civil War. He was bought up to work on the farm and received little formal education. It was said that the inspiration for his career in transport came from studying a traction engine he saw at the side of a road near Detroit. He left the farm and took on work in Detroit repairing watches, something for which he had shown an early aptitude. It was whilst in this job that he heard of Nicolaus Otto's invention of an internal combustion engine. In 1891 he took a job as a steam engineer with the Edison Company in Detroit, but he already had his eyes set on vehicles powered by internal combustion.

His first attempt was a powered bicycle, but the engine proved too powerful, leading him, in May 1896, to produce his first car. It was engineered from whatever pieces of metal he could lay his hands on – it had two pairs of wheels and (originally) a saddle taken from a bicycle, and the cylinders were machined from a steam engine's exhaust pipes. There were two forward gears, no reverse, and no brakes to speak of – to stop you had to go through a process of disengaging the clutch before applying a footbrake which, in an emergency, would be the equivalent of having no brakes at all. Perhaps mindful of this, Ford fitted his prototype with an electric bell to warn other road users of its approach. More immediately, the car was too big to get out of the doorway of the house, so a brick wall had to be demolished to enable him to give it a first test drive – much to the delight of his landlord.

By the time Ford was designing his second car, his employers at Edison gave him the ultimatum of either working for them or working on his pet projects. So Ford left Edison and found backers willing to put up $10,000 for the development of his car. However, Ford's determination to perfect his product outreached his budget, and, as expenditure passed the $86,000 mark without any start to manufacturing, he and his backers parted company.

Ford increased his profile considerably by building racing cars. The publicity helped him to raise new backers, who invested a further $60,000. Once again Ford's demanding standards and his preoccupation with racers, rather than production cars, led to clashes with his financial backers. Another formidable engineer and entrepreneur, Henry M. Leland, was brought in to bring some focus to the business. Predictably, he and Ford fell out, and Ford quit the company. The backers were going to liquidate the business but, as we saw, Leland persuaded them to continue into production with a version of a Ford design, using one of Leland's single-cylinder motors. Thus Ford, improbably, gave birth to Cadillac.

For his third attempt to become a car manufacturer Ford took advice from some successful Detroit businessmen, ranging from a banker to a coal merchant, and, in June 1903, the Ford Motor Company was finally launched. The Dodge brothers, who ran a large machine works in Detroit, undertook to supply Ford with engines and running gear in return for a 10 per cent share in the company. For investors large and small the company was to prove a goldmine. A

Henry Ford with his racing car *999*. At the wheel is a youthful Barney Oldfield – one of the few photographs of him without a cigar clenched between his teeth.

banker, John T. Gray, saw his $10,500 investment turn into $26,250,000 within sixteen years, and a Miss Rosetta V. Couzens would eventually receive a total income of $355,000 from the modest $100 share she bought.

His first car to go to production was the 1903 two-cylinder 8hp Model A (originally marketed as a 'Fordmobile'). At this stage Ford had not discovered the magic formula for cutting production costs. At $850 his Model A cost $100 more than the equivalent (and virtually identical) Cadillac. Nonetheless, within a year he was turning out 350 of them a month, his business built mainly on good after-sales service rather than price. He followed this up with the Model B, which proved to be too big and costly to sell in large numbers. A new factory opened in 1904 enabling him to increase production to 100 cars a day by 1906. His new bestseller was the Model N, the price of which he had by then managed to reduce to $500. For some years there were disputes within the company between those who wanted to build for the mass market and those who wanted to go up-market. This was only ended when Ford was able to buy out the 'big car' supporters in 1906/07. By that time 8,500 of his Model N had been produced, the company had made over $1 million net profit and Ford had become the world's largest car manufacturer.

But it was the Model T, first launched in 1908, with which Ford will forever be associated. It was not especially cheap when first launched ($850), nor was it particularly comfortable or easy to start. It also had some odd quirks, described later, but it was lighter, more rugged and more powerful than the opposition, as well as being simple to maintain. The combination of lightness and strength came as a result of Ford picking up a piece of light but strong steel left by a French car at the site of a road accident which he had analysed and found that it contained vanadium. This form of steel was not then generally available in the United States, and Ford's use of it enabled him to build the robust Model T with a weight of less than eleven-hundredweight.

The Model T continued Ford's use of racing as a means of promoting his cars (despite its modest performance in standard form, there was a thriving Model T racing fraternity), but it became better known as the car that opened up the west. The philanthropist Robert Guggenheim put up a prize for a trans-America race from New York to Seattle. Two Model Ts were entered against much more expensive European opposition, including a 60hp Itala. But in the trackless wastes that made up much of America beyond the eastern seaboard, it was the Model Ts that triumphed, coming first and third. The winning car completed the 4,106 miles in twenty days and fifty-two minutes, seventeen hours ahead of its nearest rival. The Model T can claim much of the credit for the development of the American road network in the years that followed.

Just as Harry Lawson tried to create a monopoly of car production in Great Britain, so America had its own monopolist who, for a time, enjoyed considerable success. George B. Selden was a patent attorney from Rochester, New York. In 1879 he lodged a patent application for a form of two-stroke engine. By means of lodging over 100 subsequent amendments to the original patent (many of them plagiarised) he managed to delay its final registration until 1895. By this time the first hints of a booming American car industry were beginning to emerge, giving Selden a potential money-spinner. His aggressive protection of his patents led many of the fledgling companies to conclude that co-operation was better than defiance, and they allied with Selden to form the Association of Licensed Automobile Manufacturers (ALAM). They paid royalties on every car they made or imported, but also had a share in the royalties paid by newcomers to the industry. More important than this, they could control the volume each company produced, and even whether they could manufacture at all.

Ford decided to confront the dubious content of Selden's patent head on, leading to a legal battle that ran on for six years. In parallel to the court case, there was an advertising battle between Ford and ALAM, who published lists of the lawsuits they had initiated against patent infringers, threatening to sue every owner of a Ford individually. 'Don't buy a lawsuit with your car', their advertisements warned. Ford, for his part, promised in his advertisements to assume responsibility for any action the Trust might take against individual owners. The case dragged on until 1911 and cost Ford huge amounts of money. He initially lost the case, but was eventually successful on appeal, the court deciding that because Selden's patent related to a two-stroke motor it was fundamentally different to Ford's system. In addition to the freedom to manufacture, he won huge public support for taking on the monopolist in this way.

FORD AND MASS PRODUCTION

The working conditions there (at Fords) were so intolerable that anyone with a choice preferred to work elsewhere.

(Brandon, page 113)

At the same time as fighting Selden in the courts, Ford was planning the move that would revolutionise car building in particular and industry in general. Many of the earliest car engines were machined using quite primitive tools, and most of their finishing was done by hand. Chisels, files and scrapers were used to get the various parts to fit together, with the result that each engine was hand-crafted and standard replacement parts for it could not be provided off the shelf.

This was not an option for Ford because he could not find the number of skilled mechanics needed for this traditional method of car manufacture. Instead, his approach was to install sophisticated machinery that would machine parts with such accuracy that unskilled men could assemble them. His new Highland Park factory began producing in 1910, and by 1913 the five key principles of mass production were in place:

ACCURACY – the interchangeability of parts, aimed at minimising the amount of hand-fitting needed, both when the car was being made and when it was subsequently serviced;

CONTINUITY – of the manufacturing process (i.e. the production line);

SYSTEM – careful programming and control of all the moving lines to avoid a breakdown stopping the entire process combined with supervision of employee working methods, controlled supply of raw materials, a parts inventory and product shipping;

SPEED – the use of the latest machine tools and the sub-division of tasks into the simplest possible steps; and

ECONOMY – the replacement of costly forgings by quality castings.

The interchangeability of parts was not an insight unique to Ford. Lanchester used the approach in manufacturing his cars (though his backers despaired when it took him two years to design the machinery to make the interchangeable parts). Ransom Olds introduced many of the principles of mass production to his Oldsmobile factory before Henry Ford. At Brooklands in 1908 Cadillac gave a well-publicised demonstration of the principle of interchangeability. In this three of their production models were stripped down to their smallest individual components (over 2,000 of them), and those components were shuffled about and then re-assembled, or exchanged with standard off-the-shelf spares. The cars were reassembled using only basic tools such as screwdrivers, pliers and wrenches, and were found to function perfectly.

A Model T chassis being tested at Ford's massive Highland Park factory in 1914.

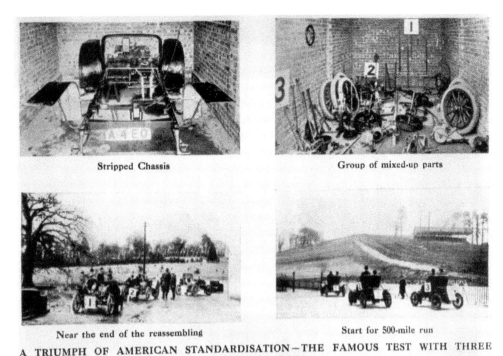

Stripped Chassis   Group of mixed-up parts

Near the end of the reassembling   Start for 500-mile run

A TRIUMPH OF AMERICAN STANDARDISATION – THE FAMOUS TEST WITH THREE CADILLAC CARS

Photographs of the publicity stunt to demonstrate the consistency of Cadillac's manufacturing.

The moving production line was another central part of the process, but again it was not a new idea. Ford probably saw it in the meat-processing plants of Chicago, but more than four centuries before that the Venetians had used a similar system for re-provisioning their galleys. They would tow them down a canal, lined with rows of warehouses containing all the different types of provisions they required, from food and arms to rigging and even a fresh crew. These were loaded as the galley passed and the vessel left the canal ready for sea.

Two other factors worked in favour of Ford's mass production. The first was having a domestic market big enough and un-stratified enough to warrant the considerable investment a production line like Ford's needed, something that did not exist in England until much later. The second was relative stability in car design: in the very earliest years of the motor car era the technology was evolving with such speed that even cutting edge models could become virtually obsolete within a handful of years. In the sort of time Lanchester took to gear up for his production, a car ran a serious risk of becoming obsolete. But by 1908 car design was sufficiently advanced for the Model T to remain a marketable proposition for twenty years (albeit with some improvements over that period, and the fact that it sold on price and reliability rather than advanced engineering, at least for the latter half of its life).

Ford controlled every part of the production process. He owned the mines producing his iron ore, the forests growing his timber and the plantations producing the rubber for his tyres. His production techniques radically reduced the labour time required to assemble a car: a chassis that had required twelve-and-a-half man-hours to build could now be completed in one hour and thirty-three minutes. Costs were correspondingly reduced. A Model T roadster that had cost $825 in 1909 cost $260 by 1925.

As his costs went down his sales went up, finally reaching a peak of over two million in 1925. In our pre-war period his sales figures were more than five times those of his nearest rival, reaching a peak of 308,162 in 1914. At one point Model Ts represented a half of all the cars in existence.

Ford ploughed large parts of his profits ($10 million in 1914) into rewarding his workforce (or, depending upon your point of view, bribing them to accept working conditions of great discomfort and mind-numbing tedium). In 1914 he raised the minimum wage for his employees to $5 a day (at a time when the average American factory hand received just $11 a week). Fellow businessmen and investors feared for Ford's sanity, but many in his workforce saw him as the new Messiah. This did not prevent the man and his methods coming in for public criticism, in films such as Fritz Lang's *Metropolis* of 1927 and Charlie Chaplin's *Modern Times* of 1936, and books such as Aldous Huxley's *Brave New World* of 1932.

Ford's decision to pay his workers this unprecedented rate was not born of pure benevolence. The tedium of the assembly line was having alarming effects on labour turnover, which at Highland Park in 1913 was running at no less than 370 per cent a year, with 71 per cent of employees lasting less than five days. He had to hire 963 new employees to secure 100 who would stay the course. With each new member of staff costing $100 to train, staff turnover was costing Ford $3 million a year. The new wage deal reduced staff turnover by eight-fold and also produced significant reductions in absenteeism. It also produced near-riots on the day of its announcement, as thousands of would-be employees assembled outside the plant. Ford's staff eventually had to turn the hoses on them to calm them down. Even so, Ford was hardly able to trawl the sunlit uplands of labour supply. An analysis of his workforce in 1914 showed that it contained high proportions of groups that would otherwise have found it difficult to compete in the jobs market, including ex-convicts, people with physical and mental handicaps and newly arrived immigrants with little or no command of English. No less than 71 per cent of his 1914 workforce had been born outside the United States, and a survey of his labour force the following year showed forty-nine different nationalities among his 18,020 employees.

From 1911 Ford also ran a very successful British operation at Trafford Park, Manchester, assembling Model Ts from imported parts. It was only a year behind the American parent plant in introducing Ford's new production techniques. He paid just £2,000 for a long lease on a five-and-a-half-acre site in the new industrial park, and employed Percival Perry, a former associate of Harry Lawson, to run it for him. Within three years Perry had created the most efficient car-making plant in the country, with a network of around 1,000 dealers. In the year that he set up in the United Kingdom, the entire British car industry turned out 19,000 vehicles. Within two years Ford's UK production was up to 7,310 cars compared with the largest individual British manufacturer, Wolseley, who made just 3,000 in 1913.

In practice there was less than met the eye with Ford's $5 figure. Part of it was made up of a non-guaranteed profit-sharing element. But the round figure made eminently good public relations among his labour force, just as it sent shivers down capitalists' spines. Nor again was the idea his own. It was originally introduced in the British Trafford Park plant where Percival Perry offered his workers £3 a week, well above the starvation wages paid by other local manufacturers. It was Perry who related to Ford that productivity improved enough as a result of this increase in wage for it to make economic sense.

EARLY AMERICAN RACING

As we have seen, the early development of the car in America lagged behind that of Europe and was mainly confined to the east coast (due not least to the lack of suitable roads for them elsewhere). The so-called Automobile Club of America initially drew most of its membership from New York. The rules against road racing were as strict in America as in Britain (with the added twist that some country sheriffs thought nothing of using a shotgun to take out the tyres of speeding motorists who failed to slow down). This led to early experiments with car racing on trotting courses in Morris Park, New York, and Brighton Beach, next to Coney Island. Races of up to twenty-four hours and over 1,000 miles in length were staged, though tracks designed for trotting horses tended to break up badly under the wheels of cars circulating at speeds of up to 50mph.

What may have been the first road race on American soil took place in Chicago on 28 November 1895. Given the climate of the area, this choice of date and venue may be thought of as optimistic in the extreme, and, sure enough, the day dawned in the aftermath of a blizzard, with snow-covered roads. There were, however, several reasons for the choice of Chicago as the venue. First, it was a hub for national transportation networks, which made it easy for competitors to ship their vehicles to the city from every corner of the nation – despite the fact that Henry Ford could not find anybody to sponsor him to ship his car to Chicago for the contest. Second, Chicago was suffering particularly badly from the problems associated with transport based on horses (such as dung and the presence of dead horses on the streets – matters discussed elsewhere in the book). The city was also associated with the Columbian Exposition of 1893 which highlighted new forms of transport that would render the horse obsolete (the bicycle was at this time seen as much more important than the car as the successor of the horse). Third, there was the aggressive sponsorship of the *Chicago Times Herald*, whose proprietor, H.A. Kohlstaat, had heard about the 1894 Paris–Rouen event and wished to replicate it in America.

They offered a $5,000 prize for what was originally to have been a 100-mile race between Chicago and Milwaukee, Wisconsin, but this was later reduced in distance to fifty-three miles, ending at Evanston, Illinois. At that time there were only 300 cars in the whole of the United States, of which just a handful were home-grown, but eighty-one of them initially entered. The race was originally to have been held on 2 November but, as the day approached, most of the teams who had initially expressed an interest melted away. In most cases this was because their experimental vehicles were not ready in time, though many of those that were entered failed the preliminary test of roadworthiness set by the organisers. By 2 November only two entrants were ready to run, so the organisers re-classified it as a two-car 'exhibition race' (and cut the prize money to $500). The main event was pushed back to 28 November.

Only 500 people turned out to witness the start of the exhibition race – as the *Times Herald*'s rival paper, the *Tribune*, cheerfully pointed out, some 499,500 short of the *Times Herald*'s forecasted turnout – and most of them were curious commuters passing by on their way to work. Rival reporters also revelled in the fact that the cars seemed to struggle with the smallest incline, and could be overtaken by reporters covering the race on tandems. The result of the exhibition race lived down to the expectations of the event as a whole. It turned out not to be a race at all, in that only one car finished – a Benz, driven by Oscar Mueller of Decatur, Illinois. His average speed was just 6mph. The other entrant, a Dureya, suffered steering problems after hitting a team of horses and had to retire. This latter entrant had been subject to some more than minor modifications in the preparation for the race, including converting the engine from two-stroke to four-stroke to improve its power and smoothness of running.

The race on 28 November attracted even less interest – only about sixty spectators showed up for the start, which clashed with a local derby in the nearby football stadium. Many of those who did watch were not obvious converts to the cause of motoring, derisively shouting 'get a horse!' as the competitors variously crashed into things, lost wheels and broke down. The finish was even less well attended, with the estimated attendance varying between 'not fifty people' (the sponsor's estimate) and 'two reporters' (that of a rival paper). The low speed of the entrants was just as well since most of them had little or no means of stopping. One, Jerry O'Connor, driving a Benz sponsored by Macy's department store, managed to hit no less than three other vehicles before reaching the city limits. The weather was also a trial: after several hours of being exposed to the freezing elements, Oscar Mueller passed out at the wheel of his Benz. One of the race judges, a man named Charles King, was travelling with him, and he seized the controls and finished the race one-handed, the other being used to hold the still-unconscious driver in place beside him in his seat.

Because the cars were all hand-built, any spare parts needed during the race had to be fabricated en route, and Dureya's experience in the official race of 28 November illustrates the ingenuity these early motorists needed. First he broke a steering arm, so he took it to a blacksmith's shop and had a replacement fabricated. Then the igniter on his engine started to fail, so he drove it to a tinsmith's shop and, finding it closed, drove to the tinsmith's house, roused him and took him back to his workplace to complete the repair. With this done, he was back on the road within fifty-five minutes and, despite these delays, was able to win the race. But several of the entrants suffered fundamental flaws that were beyond any amount of ingenuity. A.C. Ames' steam-powered car was concocted from two bicycle frames and the body of a sleigh, but had a range of little more than 100ft. The two electric car entrants also ran out of power before reaching the city limits.

Far from stimulating interest in motoring in America, these events seemed almost to set the cause back. Notwithstanding the evidence, the *Chicago Times Herald* concluded that:

> Persons who are inclined to decry the development of the horseless carriage will be forced to recognise it as an admitted mechanical achievement, highly adapted to some of the most urgent needs of our civilisation.

But then, they were the race sponsors.

The race got little media coverage at the time, but it did lead to the establishment of the American Motor League, claimed to be America's, and the world's, first body to represent motoring interests. And despite this inauspicious start to racing in America, various other events were launched, leading to the Championship Auto Racing Team (CART) being instituted in 1902. Another eventual offshoot of this first race was the Chicago Auto Show, beginning in 1901.

Some progress was being made by 1901 in the development of the American racing automobile. Henry Ford decided to go into competition to promote his brand, and as a way of raising funds for his business. He built a two-cylinder 8.1-litre racer to take on what was

This picture, from a 1963 Standard Oil Company magazine, is claimed to represent Barney Oldfield racing the Ford *999* at Grosse Pointe, Michigan. If so, it cannot have been in the famous race with Alexander Winton, which took place the year before the *999* was built.

then thought to be America's fastest car, built by Alexander Winton, at the horse racing track at Grosse Point, Michigan. He beat it, at an average speed of 45.8mph. The following year he sought to repeat his success with a pair of four-cylinder 15.7-litre monsters named *999* (after an express train) and the *Arrow*. They were described as being 'vicious to handle', so much so that Ford had no inclination to race them himself. In his words, 'going over Niagara Falls would have been but a pastime after a ride in one of them…The roar of those cylinders alone was enough to half-kill a man.' Instead he made the curious appointment of Berna Eli – better known as Barney – Oldfield as his driver (Oldfield's colourful career is documented later in this chapter). Ford later rebuilt the *Arrow* and set a world land speed record of 91.3mph with it on the frozen Lake St Clair.

THE VANDERBILT CUP

Oh, the degradation of such a scene! As foolish as a bullfight; as vulgar as reddening the sands in a gladiatorial contest; as revolting as bartering Christ's garments for a few pieces of silver.
(Part of the Reverend Dwight Hollis' sermon denouncing the first Vanderbilt Cup, quoted in Serrano, page 128)

Down in Florida a former sports writer named W.J. Morgan started promoting a series of 'beach runs' and 'special carnivals' in 1902 on a twenty-mile strip of hard sandy beach between Ormond and Daytona. One of the outstanding competitors at Daytona was William Vanderbilt Jr. He was a wealthy young American motoring enthusiast who had competed successfully in some of the top European road races and who at one time held the world land speed record for the flying kilometre, at 92mph. He used his influence, against a host of court orders, public hearings and injunctions, to secure an exception to the 'no road racing' rule. In October 1904 he got a thirty-mile road racing circuit delineated at the western end of Long Island. The winner received the Vanderbilt Cup, a $2,500 silver trophy, along with $25,000 prize money. A grandstand for 400 of the great and the good, from President Theodore Roosevelt and the State Governor of New York down, was erected, and everyone with a spare room or a porch with a view of the circuit cashed in on the influx of visitors.

Vanderbilt set about trying to improve on the dismal safety record of European road races. Posters warning local residents about the forthcoming race were put up:

All persons are cautioned against allowing domestic animals or fowls to be at large. Children unattended should be kept off the road. Chain your dog and lock up your fowl. To avoid danger, don't crowd into the road.

But despite his best efforts, safety was little better than in the European road races. The drivers themselves were not exactly the best advertisement for road safety: the Fiat driver William Wallace managed to run over his mechanic whilst making emergency repairs, and promptly recruited a replacement from among the onlookers. In some cases crowd control was worse than in Europe; at some of the later runnings of the event elements in the crowd came armed with wire-cutters and crowbars to remove any obstacle to a view that was literally trackside, and vendors were actually selling bamboo canes with feathers on the end designed to enable adventurous (for which, read criminally foolish) spectators to tickle the faces of the drivers as they rushed past. Anything from twenty-five to fifty thousand spectators flocked to the area, parked along the sides of the circuit and wandered about on it. Train services crossing the circuit continued uninterrupted. After the winning car crossed the line in the first year's race the crowds flocked to leave and, as crowds are wont to do at large sporting events, formed long queues for the exits. Unfortunately the roads on which they queued formed part of the circuit where the other competitors were still racing. Horrific accidents were only narrowly avoided.

Arthur Duray grabs hold of his mechanic to stop him falling out of their De Dietrich during the 1906 Vanderbilt Cup race. They went on to take third place.

So the races in both 1904 and 1906 had to be abandoned before the end due to crowd control problems. There was only one fatality in the 1904 race – the mechanic of one of the Mercedes entries – but the tabloid press printed every kind of rumour and downright lie in its sensational coverage of the event. The 1907 race was cancelled entirely due to safety concerns. Although the aim of the race was to give a boost to American manufacturers, only five of the eighteen entrants to the first race (seventeen of which started) were American-built, and European cars won the first three years' races. Only in 1908 and 1909, when there was less overseas competition, did the race produce a home-grown victor.

Safety remained a constant problem, and one of the solutions that Vanderbilt hit upon was the construction of one of America's first modern-paved purpose-built roads for cars; the Long Island Motor Parkway. This was intended not only to provide a race track but also to open up Long Island to economic development. Construction of the forty-five-mile road (originally intended to be seventy miles) began in 1907, and part of the 1908 race was run over it (the construction of the Parkway was one of the reasons the race was allowed to go ahead). The full length of the road featured in the 1909 and 1910 races, but a further serious accident in 1910 led to the New York legislature banning racing on all public roads. If the Parkway's career as a racetrack was a short one, its life as a first-class road was barely longer. By the 1920s advances in both car and highway design had left the Parkway obsolete, and it was finally closed in 1938. Parts survive today as local roads and cycleways.

The Vanderbilt Cup nonetheless became a huge popular success. A quarter of a million people were said to have come to the 1906 event. It became the American equivalent of the Gordon Bennett Trophy, and did in the end provide a stimulus to home-grown manufacturers. But by 1910 the Vanderbilt Cup provided an almost continuous trail of carnage. Several people – participants and spectators – were killed and more than twenty others ended up in hospital, seriously injured. The race – with average winning speeds of 65.18mph – had far outgrown the circuit, and it was the last of its kind to be held there. From 1914 the Vanderbilt Cup moved to Santa Monica, California, where the purpose-built new circuit could cope with high-speed spectacles, including a 'wall of death'. The first race there was won at an average speed of 75.5mph, and one of the leading protagonists was a Mercer capable of 117mph in a straight line. But America by then had another circuit that would become more famous than all the rest.

INDIANAPOLIS...

The Indianapolis 500 has been held every year since 1911, making it the world's oldest motor race with an unbroken record. The circuit was the brainchild of a property magnate called Carl G. Fisher, who made his fortune developing Miami Beach. He was also a professional racing cyclist, and in 1895 became the first man in Indianapolis, where he lived, to own a car. Within five years he had set up a business selling cars in the city. He soon saw the promotional opportunities that motor racing presented and, in 1905, entered the Gordon Bennett Cup Race in France. He was unable to compete because his car exceeded the 1,000kg weight limit, but the experience of attending the event showed him both how much public interest there was in motor racing and how difficult it was to manage such events on public roads (both in terms of safety and of charging admissions to make the event profitable).

His solution was a giant version of the velodromes he had raced on as a cyclist, two-and-a-half-miles long, with banked curves at either end of the oval circuit and grandstands from which all the action could be seen. His design team took advice from the builders of Brooklands, but instead of a concrete running surface they opted initially for asphalt. The first race on the new circuit, in August 1909, was for motorcycles, but even they caused the asphalt surface to break up. They were replaced by over three million brick paviors. Regular race meetings began to be held, but attendances began to dwindle, causing Fisher instead to concentrate on a single large-scale event each year. Originally this was to be held over twenty-four hours, but he eventually opted for a shorter 500-mile event – and the Indianapolis 500 was born. The first running in 1911 had prize money of $27,500 and attracted forty entrants and a crowd of about 80,000. The winner, Ray Harroun, was also the co-designer of his car, the 7.8-litre Marmon Wasp. He took nearly seven hours to complete the race at an average speed of 74.57mph.

...AND THE FIRST RACING DRIVER CELEBRITY

Who do you think you are? Barney Oldfield?
(Oldfield was the first racing driver whose name was popularly taken in vain in this way, when police arrested speeding motorists)

A 9,934cc Peugeot 148 dating from 1913, the year in which Peugeot won the Indianapolis 500.

Barney Oldfield (1878-1946) was born in Wauseon, Ohio, of Anglo-Irish parents. From his earliest years he was obsessed with speed, cadging rides on the town's fire engine as it raced to put out fires. From there he graduated to racing bicycles, 'borrowing' (without permission) the cycle of a resident in the Toledo Hotel in which he worked as an elevator operator. By 1896 he was earning a precarious living as a professional racing cyclist.

Despite having no experience of even driving a car, far less racing one, he applied successfully to Henry Ford in 1902 to be the driver of his thunderous 15.7-litre racing special, after Ford decided it was too much of a handful for himself to drive. Ford later had cold feet about this appointment and begged Oldfield to withdraw, but, despite this, Oldfield displayed a gung-ho spirit that positively bordered on the suicidal, saying:

> This chariot may kill me, but they will say afterwards that I was going like hell when she took me over the bank.
>
> (Hough, page 41)

He added for good measure 'I might as well be dead as dead broke.' According to some accounts Oldfield only learned how to operate the controls of the car on the morning of the first race. Others say he had the luxury of a whole week of lessons. Whichever it was, Oldfield surpassed Ford's expectations and failed to live up to his own by lapping the entire field and winning the race in record time. His cornering technique was particularly remarkable, consisting of slowing the car by sliding it sideways, speedway motorcycle-style, rather than troubling with the brakes. The following year he became the first American to cover more than a mile in a minute, and he was hired by Alexander Winton (whose car he had defeated in his first race) as his works driver.

Oldfield combined his racing career with visiting fairgrounds and exhibitions all over the country, where he would race all-comers (often in races which were fixed, in order to heighten the drama, and from which he made a great deal of money). All-comers included not just cars but aircraft, with Lincoln Beachey's Curtiss biplane being a regular opponent until it nose-dived into the track at Emeryville, California. He was also involved in record-breaking attempts, and in March 1910 he recorded a world land speed record of 131.27mph in his 21.5-litre Blitzen Benz on Daytona Beach. This further increased his box office appeal at the fairgrounds and, at the peak of his popularity, he could charge as much as $4,000 appearance money. He lived a correspondingly high-rolling lifestyle, mixing with the celebrities of the day and travelling to events in his own personal railway carriage.

His name became synonymous with speed and daring. Firestone used it to market their tyres, with the slogan '"Firestone tires are my only life insurance" says Barney Oldfield, world's greatest driver.' He starred in a Broadway musical called *The Vanderbilt Cup* for ten weeks in 1906, and in 1913 became one of the first sporting stars to play himself in a moving picture. In Mack Sennett's film *Barney Oldfield's race for a life* he raced a train to rescue the heroine, whom the villain had tied to the tracks. His involvement in the cinema, either as an actor or a technical advisor on motor racing-related films, continued until 1941, and he died close to Hollywood, in Beverly Hills, California.

Photographs of Oldfield show a pugnacious-looking man with a cigar permanently clamped in the side of his mouth, and he lived up to his image, tending to get involved in bar room brawls. His dubious fairground activities led to him being banned from official racing events by the American Automobile Association. He also exhibited some of the racist tendencies that were more commonplace in his day. After the African-American boxer Jack Johnson took the world heavyweight championship from Oldfield's friend Jim Jeffries, Oldfield took on Johnson in a motor race, allegedly to demonstrate the superiority of the white man.

Oldfield did, however, manage to get himself unbarred from official events for long enough to enter the Indianapolis 500 in 1914 and 1916. He only finished fifth on both occasions, but became the first person to record a 100mph lap of the circuit. He retired from competition in 1918.

CHAPTER 10

POOP! POOP!
ON BEING AN EARLY
MOTORIST

'Glorious stirring sight!' murmured Toad, never offering to move. 'The poetry of motion! The real way to travel! The only way to travel! Here today – in next week tomorrow! Villages skipped, towns and cities jumped – always somebody else's horizon! O bliss! O poop-poop! O my! O my!'

(Kenneth Grahame, *The Wind in the Willows*, 1908)

To rush through the air at the speed of a torpedo-boat destroyer, down a narrow curving road enclosed with hedges and without being able to see what was in front of us, was a novel and thrilling experience. One minute we were 500ft above sea level, and the next only 300. We had accomplished this rapid descent of 200ft in only a few seconds of breathless suspense when the slightest error of steering would have landed us into one bank or the other or plunged us into the midst of cyclists who were waiting at the bottom of the hill to see how we should take the awkward bend. We did it magnificently and all the while our engine was actively propelling us onwards and thus adding to the velocity which had been imparted to the vehicle by its momentum.

(Journalist for the *Automotor Journal*, accompanying one of the 1896 London–Brighton entrants)

In this chapter I have tried to recapture something of the experience – the exhilaration, the frustration and the sheer hard work – of being one of the early motorists. As we saw at the start of the book, the earliest motorists were as strange to some communities as beings from another planet, and were likely to be closely questioned about their exotic chariots.

But the public relations part of a motorist's role was positively pleasurable compared with the sheer hard work of getting from A to B – particularly if you made the wrong choice of vehicle. Elsewhere in the book I gave an account of the fraudster E.J. Pennington's malign influence on the British car industry, and referred to a nightmare journey H. W. Egerton tried to make in one of Pennington's appalling cars from Manchester to London. Egerton recorded part of this ordeal in a letter to his mother. It was quoted at length by Nixon and bears repeating here, as an illustration of the very worst experiences that being a pioneer motorist could involve.

July 4, 1899

Having wired up the machine, with a fresh battery, I found the sparking plug wire was leaking, and so sparking through the insulation and giving me violent shocks. Having joined the wire the trouble ceased, and I left the Swan Hotel at 11 a.m. The engine ran well at first and then began to sulk which made matters very difficult on some of the hills I had to climb. I had to walk up most of them holding the clutch of the low gear with one hand and the belt shifter with the other while I steered with my body.

In the middle of the first hill, the belt suddenly ceased to bite, and instantly the car ran backwards. I promptly put it into the hedge. Having spent much time cutting and refastening the belt and adjusting the chain to suit, I managed to get up the hill. The next trouble was a dead sparking plug. Then the belt gave more trouble and had to be cut again and refastened. I had to walk up nearly every hill and sometimes had to push as well.

At 3 p.m. while running free down a steep hill with a corner at the bottom, off came the chain and away went the car at a frantic speed. As I foresaw all along, the steering is absolutely dangerous.

Having resisted the temptation to abandon ship until the corner was turned, I got the car in hand once more and went back on foot for the chain, which I found lying in three pieces. Its condition was hopeless, so I fitted a fresh one from my 'spares department' and after taking off three links and re-riveting it up again, I was just about to start when I discovered that the rubber pipe carrying the water had started a big leak and had thoroughly saturated the belt with boiling water, which, with the grease already on it, most effectively prevented the engine from pulling the car on the level and with low gear engaged.

After cutting it again, it still refused to bite, and there I was, four miles from Lichfield at the bottom of one of the four steep hills separating me from the town; time 5 p.m.; last meal (breakfast) 8.15 a.m.

Realising the utter hopelessness of trying to run a car with a belt like elastic, I took a short rest and thought matters over, but time was getting on so I set to again. All my dodges ended in failure, however, and there I stuck – the engine going like an express train and the car placidly ignoring the fact that there was an engine within miles.

Then came a mighty brewer's dray driven by a jovial drayman. We hitched the car on behind and covered fully one hundred yards when the accident list increased; bang went the rope.

Having made fast a cart rope tested to bear quite three tons, we tried again, and in the course of an hour and a half we arrived in triumph and absolutely safe at the Swan Inn, Lichfield, having accomplished no less than sixteen miles in nine hours.

At the present rate of progress, it would take me three weeks to reach London, but I realise of course that I cannot always expect to cover sixteen miles in nine hours.

(Nixon, pages 136-7).

Even if the car was working as intended, the progress of some of these vehicles was hardly the last word on comfort. Booth Tarkington recalled a 1903 trip through Belgium in the *Saturday Evening Post*:

The large wooden wheels had solid rubber tyres, and their passage over an ancient, stone paved road would have been stimulating to the spinal ganglia if the performance of the engine's two large cylinders had not already attended to that... Horseless vehicles were intended evidently for people with rubber backbones and no fretful imaginations.

(Pettifer, page 45)

STARTING AND STOPPING

Just the basics of starting and stopping early vehicles involved an amount of work for the pioneer motorist that is unimaginable for their modern counterparts. Taking starting first: because the earliest cars were prone to breaking down, their drivers frequently had to re-start them. This was no easy task, as Ruiz explains:

> [It] involved first lubricating the connecting rod and the crankshaft by filling the special grease points, checking the cylinder oil-feed levels and lubricating the transmission. Then, at least until the constant-level type invented by Bernardi and perfected by Maybach in 1895 became more popular, the driver had to fill the carburettor with exactly the right amount of petrol. It was then possible to fire the burners to make the platinum tubes incandescent, an operation which took some time; alternatively, if the car had electric ignition (initially fitted only on De Dions and Delahayes), the driver had to check the wires, which were particularly prone to short circuit, and the condition of the battery. Finally, with the exhaust valves open to avoid undue high compression and with the clutch engaged, the driver could crank the starting handle which acted on the flywheel. As soon as the engine started to turn, the driver closed the exhaust valves, cleaned his hands and took his position at the wheel.
>
> (Ruiz, page 18)

Cranking a car with a starting handle was a hard, uncertain and sometimes dangerous means of getting it started. Broken wrists and worse were not uncommon among drivers trying to crank their cars. When the Duke of Bedford found himself stranded miles from home after his chauffeur broke his wrist while starting, he took to being accompanied by two chauffeurs, in case it happened again (which must have been comforting for the chauffeurs). With the Benz Viktoria there was not even a starting handle, and Jellinek-Mercedes describes the alarming starting procedure for this model as follows:

> While a courageous helper turned the flywheel, another had to hold onto him, so that a sudden forcible backfire did not throw him against the vehicle.
>
> (Jellinek-Mercedes, page 45)

Various forms of self-starter were tried, with the Arnold Motor Carriage Company claiming to have been the first to apply one in 1896-98, with Mors introducing one in 1905. Curiously enough the electric models were initially the least successful – compressed air versions looked like a much more promising avenue, and there was even a clockwork variety offered for sale in 1911. But in that same year Charles Kettering invented a workable electric self-starter for petrol cars. This invention coincided with Byron Carter, a friend of the head of Cadillac, Henry Leland, having a fatal accident. Carter went to the aid of a lady motorist who was attempting to hand-crank a car. The car backfired on him so violently that it caused him to break his jaw, an injury from which he contracted gangrene and died. Henry Leland subsequently became the first person to introduce the electric starter as standard on his 1912 Cadillac range, giving them great appeal to female motorists, and to a more up-market clientele who did not care to soil their hands (or do themselves an injury) by way of hand-cranking their cars. They caught on rapidly with the American public, and within a couple of years it was almost inconceivable to offer them a new car that did not have an electric starter. Some manufacturers stupidly located their starters where the starting handle used to be. This meant that, in the event of a starter motor failure, the only way of getting the car going was to tow or push-start it, or to jack it up and spin the back wheels with the engine in gear.

As for stopping:

> Since the brakes only worked in forward motion, a stone was always carried and placed immediately behind a wheel if the car stopped on a hill: however, some cars, such as the Panhard-Levassor, had a special prong – then known as a sprag' or 'bequille' – which was dropped when needed.
>
> (Ruiz, page 16)

Many of the early cars had rear-wheel braking only. Crossley was one manufacturer who introduced four-wheel braking as standard in 1909, only to abandon it in 1912. Even Grand Prix cars, by the outbreak of the First World War, were shy of this innovation. Their fatal error was to fail to connect all four wheels to the foot brakes, instead linking the rear brakes to a lever at the side of the car. Cecil Bianchi, Crossley's works manager, recalled that this led to:

> Locking of front wheels and loss of steering, the whole load of braking being taken on the front axle and springs, which gave endless bother in broken springs, twisted axle beams and stub axle failures, even with radius rods fitted.
>
> (Montagu, page 63)

There were alternatives to stopping with the brakes, as the *Chauffeur's Companion*, published in 1909, suggested. But they were not for the faint-hearted:

> If…on a fairly broad and slippery road you have the alternative of going on and killing someone, or stopping, and you find you cannot stop, there is one desperate remedy which

The 1912 Cadillac, the first car to be fitted with an electric self-starter.

may or may not come off, and that is to try and swing the car round with the bonnet pointing to where the back was. To do this (and I only advise it in cases of absolutely dire necessity) swing your steering wheel hard round and at the same time open your throttle and jam in your clutch with a jerk. This should skid the car round and when you are round you must get the wheel over on the other lock, and send her straight along the road again. The impetus of the car forcing her way in the original direction and the engine pushing her the reverse way (when you have got her round) will neutralise each other and she should stop almost dead.

(E.S. Turner, page 153)

RE-FUELLING

I described elsewhere the difficulty encountered by the Bollee brothers in finding fuel when they arrived in London for the first London–Brighton run. Even three years later, when *The Autocar* published a list of petrol stockists for motorists, they were only able to find four in London and twenty-nine elsewhere. The use of paraffin for lighting had created a substantial oil industry. The refineries of the day broke the crude oil down into three parts. The heaviest part was used for lubricating oils. The middle part was the paraffin. The lighter part, naphtha, was more or less a waste by-product, and in some senses a rather inconvenient one in that it was highly volatile and explosive. In these early days the petrol could be of a rather variable quality, so motorists used to test its specific gravity which, by European standards, was supposed to be around 0.68 (the Americans used gasoline, whose specific gravity was 0.66, and a slightly heavier product, Benzoline, with a specific gravity of 0.72 which could also be used when petrol was not available). Standing for any length of time could also cause the fuel to lose its volatility, making the car difficult to start.

Incidentally, Benzene was only known as petrol from about 1900 onwards, the latter name apparently invented by Frederick Simms, the founder of Daimler UK, and coined as a trade-name by a company called Carless, Capel and Leonard, who imported it.

But the legalisation of motoring did not solve the problem of the availability of fuel overnight. In the days before petrol stations any long journey had to be planned like a military operation. Nixon gives us an example of his failure to do so, having made a journey from Putney to Margate on a motorised tricycle in the summer of 1900:

After spending a few days there, [I] prepared for the return journey. I tried to get some petrol, but it was not sold in Margate. I tried every cycle shop in the place, and not only had they none, but no one could tell me where I would be able to buy some, either in Margate or in any nearby town. I purchased as many bottles of benzine at chemists' shops as I was able, but I could not get sufficient to carry me home. After I had visited all the chemists' shops, I doubt whether Margate could have produced more than an eggcupful of benzene. I was at my wits' end to know what to do for the best and was on the point of coming back to London by train so that I could take a small supply back to Margate, when I happened to remember that H.J. Lawson of the British Motor Syndicate had a house there; I called on him and he was good enough to let me have a couple of gallons.

(Nixon, page xv)

The market for petrol stations at this time may be judged by the fact that, on his way back from Margate to Putney, Nixon encountered just one other motor vehicle. The motorist was therefore left with two choices; to buy it from a chemist, who sold it in tiny bottles for use as a dry cleaner, or to buy it in bulk (quantities of almost 300 gallons at a time) in wooden barrels, from which much of the fuel would evaporate before they got around to using it. The only sources of petrol in sensible quantities were the refineries or the motor manufacturers themselves, neither of which were necessarily on everybody's doorstep. In recognition of this problem, the 1896 Act allowed the private motorist to keep their own individual stores of up to sixty gallons of petrol without needing a licence. The Act required that 'petroleum should not be handled or tanks filled by artificial light, unless such light is so arranged or protected that it shall not cause an explosion.'

In Britain the first petrol filling station did not appear until 1919, when the Automobile Association opened an experimental one at Aldermaston in Berkshire. They built seven more, but moved out of petrol retailing once private enterprise got involved. Prior to that, from 1904 the RAC issued a list of approved hotels whose services included holding a store of petrol, or being willing to fetch some on request. In America private enterprise took hold rather sooner and, before the First World War, farmers were setting up impromptu roadside petrol stations involving no more than a drum of petrol and a wooden sign.

ROADSIDE REPAIRS

If you desire to travel fast,
A motor car is unsurpassed;
Should you desire to travel far,
Trust not too much a motor car.
(Part of a poem in *Punch Magazine*)

If petrol was hard to come by, spare parts for the cars themselves were even rarer. Improvisation was the order of the day and the earliest drivers also had to be competent (and inventive) mechanics. As we saw earlier, Frau Benz used an item of her underwear to repair an electrical fault, and Nixon gives a delightful account of how he also resorted to ladies' underwear for

Few garages or petrol stations existed before the First World War. This photograph of Lawrences' of Brixton dates from about 1924.

a running repair. In the summer of 1902 he was driving an early Benz from Eastbourne to Brighton when the trembler (a piece of flexible steel in the distributor) broke. He was able to improvise a temporary roadside repair that just about limped him into Lewes, where he could think about a more lasting remedy. The only item he could think of that contained a piece of spring steel of about the right dimensions was one of the ladies' corsets of the day. (As an irrelevant aside, you may wish to know that one of the original manufacturing lines for the Peugeot Company, before they went into making cars, was precisely these metal strips for corsets.) So Nixon found a shop in Lewes selling ladies' underwear. The arrival of a young man driving a horseless carriage caused stir enough, but when he asked the assistant to sell him a lady's corset, the entire shop was agog.

'What size?' the shop assistant asked.

'It matters not two hoots,' he replied 'so long as the busks are made of metal.' As soon as the garment was produced, he further amazed the shop by producing a pocket knife and shredding it to remove the said piece of metal.

> Any lingering doubts which might have remained in the minds of the shop assistants as regards my sanity were completely and immediately dispelled by this final act of vandalism and palpable lunacy. Here, beyond all shadow of doubt, was a harmless young imbecile at large, and what made the situation a thousand times worse, in charge of one of those new-fangled motor-cars…I was indeed fortunate to escape from the shop without inquiries being made.
>
> (Nixon, page 32)

Nicholson gives us a depressing catalogue of the shortcomings of the motor car of the mid-1890s, which were likely to mean the owner would spend more time under the bonnet than in the seat. Starting with the engine, wind, rain or bumps in the road could put the hot tube ignition burners out. If they came into contact with petrol vapour the engine might catch fire. If they ran too cold you suffered from late ignition and loss of power, while too much heat meant pre-ignition, overheating and backfiring, added to which the tubes themselves were prone to breaking. But even so, they were still more reliable than the battery and coil ignition of the day, which was forever short-circuiting or suffering from accumulators that mysteriously discharged themselves or simply fell to pieces.

A combination of shortcomings regarding available raw materials of the day and primitive technology were responsible for the problems with the inlet valves, most of which operated off the vacuum created by the induction stroke of the engine. In order to do so, they and their springs had to be very light, with the result that they broke at regular intervals. They also tended to get gummed up or knocked out of shape, both of which resulted in a loss of compression and power. On a car in regular use, the valves may have needed to be ground in once a week – a job which Nicholson says took two to three hours. Also, leaking carburettor floats frequently starved the engine of petrol.

Inadequate materials and technology were also to blame for the warping of cylinder heads. Early manufacturers had great difficulty maintaining a gas-tight fit between them and the engine block. This led to a whole host of troubles; a loss of compression, hot gases from the engine getting into the cooling water causing the car to overheat and water getting into the cylinders. A warped cylinder head usually meant having to get it machined to make it fit again.

Chassis construction was still stuck in the days of the horse and carriage, with the heavier cars relying on wooden frames with iron plate reinforcements. Whilst these may have worked for carriages they were not up to the stresses imposed by motoring. They flexed, in turn putting impossible stresses on the transmission and causing gear bearings to wear out and the clutch either to slip or to develop a bite like a rottweiler. Until Panhard became the first to introduce the enclosed gearbox to their production cars in 1896, wear in that department was compounded by the mud and grit that was constantly being thrown into it. As for the solid rubber tyres, they may never have punctured, but they were prone to falling off at high speed.

For the inexperienced motorist, which in those days was almost everybody, guidance was to be found in the 1902 *Badminton Library of Sports and Pastimes*, where no less a celebrity than the Hon. Charles Rolls contributed a piece called *The caprices of the petrol motor* (reprinted in Hough, pages 22-36). This explained to the uninitiated the possible reasons why their motor car might not be working. Rolls found it necessary to begin this fourteen-page catalogue of woe with a disclaimer:

> …I have been compelled to enumerate a very long and somewhat formidable list of evils, and lest a glance at this should frighten off any would-be motorist, it must be clearly understood that the list comprises possibilities as distinct from probabilities…Many cars have been run by amateurs for thousands of miles without the occurrence of any trouble.

I will mention just a few of Rolls' techniques that may be less familiar to the modern motorist. For example, poor starting in cold weather could be due to the failure of the petrol to vaporise properly, and he offered a number of means of warming it up, the most radical being:

> In obstinate cases … to warm the carburettor underneath by means of a little cotton waste soaked with methylated spirit. There may be a slight flare up, but this will assist the carburetion, and there is nothing to fear from it so long as the main supply cock has been carefully closed.

In the days when petrol had to be obtained wherever you could find it, quality control was not always a strong point, and your fuel could lie at the heart of your problem. A good clue was if there were things floating in it, and Rolls advised that, 'petrol should never be poured into a tank, except through a funnel fitted with a very fine gauze strainer or a piece of muslin.'

But even once it was safely in your tank the petrol of the day could go off. According to Rolls it could become 'heavy', especially if it had had free access to air. An instrument called a densitometer was available for measuring the specific gravity of petrol. The best specific gravity for starting was 0.68. If your fuel showed up heavier than 0.70 it should either be changed or your tank shaken up.

Some starting problems could be solved with nothing more than a little elbow grease. If you were not turning over the starting handle fast enough than the petrol would not vaporise properly, so Rolls advised that 'a few smart turns with the whole of your energy will be more likely to start an engine than hours of slow grinding.'

Other problems could be the symptoms of more serious faults: a loss of compression, for instance, could indicate a cracked cylinder head. Rolls told you how to find it:

> Hold a lighted taper over the cylinder in different positions, whilst the engine is being turned slowly; the flame will be turned on one side on meeting the leak; or soap and water may be painted about the cylinder head and bubbles should soon indicate the presence of the leakage.

But problems like a leaking radiator leading to over-heating did not deter the rugged pioneer motorist:

> If all the water has been lost on the road through the breakage of a pipe or the opening of a plug or tap, or loosening of a joint, and no water is near you, you can continue your journey spasmodically by allowing the engine to cool down, then run a mile or two with the bonnet off or open until it shows signs of overheating again, then stop, paraffin your cylinders and wait another half an hour.

Sometimes Rolls' playfulness took him into the realms of misleading and unhelpful advice to his fellow motorist. It is to be hoped that none of his more earnest readers took the following advice literally when trying to find a cylinder that was failing to fire. It also serves as a wonderful warning to men never to read just one line of instructions at a time:

> Endeavour first to ascertain which cylinder is the culprit. One method of doing this is to place your hand on each exhaust pipe while the engine is running. You will then get a bad burn from every one except that belonging to the faulty cylinder. A more convenient way – if electric ignition is fitted – is to stop the working of three out of four of the induction coils, changing about until you find the one that is at fault.

Rolls ended with a series of pieces of general advice, at least one of which is as true today as it was then: 'Don't spill the petrol over your clothes and then strike a match to light your pipe.'

A representative cross-section of the problems encountered by early motorists may be had from the account of the journey from John O' Groats to Land's End in 1898 with which this book started. In the days before radiators recycled cooling water regular supplies had to be taken on to replace that which had boiled away. On one stage of the trip they went sixty-five-and-a-half-miles on a single supply of cooling water, which they took to be some kind of record. Fortunately, in the days when most traffic on the roads was drawn by horses, regular roadside supplies of water were rather more easily come by than they may be today. Also, the clutch needed frequent adjustment, including the application of resin when it got wet. Wet or windy weather could extinguish the hot tube ignition system, bringing the car to a halt. Chains and other lubrication points needed regular attendance. On rest days the chains were removed and soaked in a mixture of petrol and oil. Brakes were also prone to frequent failure, running red hot or catching fire. Some of the brakes that worked directly on the tyres proved relatively useless, sometimes carving grooves in the tyres. Metal brake blocks also tended to wear out fairly quickly. Resin could be poured into the brakes to make them bite when they were in danger of failing, one problem being that it took two hours to remove the burnt-on resin residue afterwards.

On the credit side, it has to be said that after-sales service in those days could surpass anything available today. When Rudyard Kipling's Lanchester broke down, a phone call to the company had one of the Lanchester brothers travelling down to his house in person to discuss the problem. Try getting hold of a member of the Ford family today when you need them.

MOTORING FASHION

Frau Benz undertook one of the first serious runs in a motor car, and it was not long before women became a familiar sight behind the steering wheel. By 1904 cars had become sufficiently reliable and women motorists sufficiently numerous for the Ladies' Automobile Club to be formed; and by 1909 Dorothy Levett was able to publish the memorably titled *The Woman and the Car: A chatty little handbook for all women who motor or who want to motor*. Despite all this, Scott-Moncrieff's mother told him that, at this time, 'it was considered dashing to the point of being "fast" for a lady to be seen driving alone, unaccompanied by a chauffeur, or at least some form of liveried servant.' But by 1910 *Outlook* magazine went so far as to see woman and car as a spiritual entity:

> …the woman at the wheel is no allegory. Already her intuition has put her in touch with the automobile. Its delicacy of adjustment, its vagary of moods, she has come to understand as those of a sister organism, for what enthusiastic motorist does not refer to his car as 'she'…
>
> …when the average family comes into possession of an automobile, it is well worthwhile for the woman of the household to acquire the ability to run it… Once she has learned to drive it and to help her husband care for it, the family's use and pleasure in the car are increased several hundred per cent. If the head of the house is the only motorist and the services of a chauffeur

The necessity of keeping warm and dry led to some outlandish fashion choices for the early motorist. (*Punch*, 1901)

cannot be afforded, the car is probably idle three-fourths of the time. As soon as the woman makes friends with it, it becomes an indefatigable source of health and pleasure to her, her children and her friends.

(*Outlook*, April 1910)

But what to wear while driving? Scott-Moncrieff suggests that:

In the very earliest days of motoring, people just put on anything old and warm that could be easily shed when pushing the car home or trotting alongside it as, relieved of the passengers' weight, it chugged uphill. Motoring dress, as such, does not really start until the late 1890s when cars were able to maintain a steady 20mph or even greater speed.

(Scott-Moncrieff, page 234)

The main consideration was protecting oneself from the weather, since early motorists were very exposed. Many motorists preferred not to have a windscreen fitted, since being hit by deadly shards of broken glass from a shattered non-safety glass windscreen was a powerful deterrent, and they chose instead to face the elements.

We are asked to believe that this lady is dressed for racing! Madame Fraschini poses at the wheel of one of her husband's Isotta-Fraschinis during the 1908 Targia Floria. Her husband is seated beside her in the mechanic's seat.

To some extent fashion varied according to the nationality of the driver. French motorists favoured goatskin coats with the fur on the inside, whilst Americans preferred racoon skin (a favourite of sledgers of earlier times) and the British a more conservative blue Melton cloth coat with a fur lining. Not for nothing did one contemporary observer describe early car owners as looking like 'something between an overgrown goat and a door mat'.

The hard-peaked cap sported by continentals was also abandoned in Britain in favour of a cloth cap or bowler hat (the latter proving to be an excellent crash helmet, should the need arise). For the British, the continental peaked-cap became the very epitome of the hired 'motor servant', and therefore quite unsuitable for an owner-driver (the snooty Automobile Club warned against them for precisely that reason). In wet weather a heavy rubber poncho might be worn, and passengers might combine this with a rug wrapped tightly around the legs (drivers could try the same, but there was a very real risk of it becoming equally tightly wrapped around the pedals, with dire consequences). Some drivers even wrapped themselves in layers of brown paper underneath their coats.

Turning specifically to ladies' fashions, and returning to the *Badminton Library of Sports and Pastimes* of 1902, Lady Jeune, wife of a leading lawyer and something of a fashion icon of her day, dispensed advice on motoring dress for ladies. Her guidance could be summarised as *abandon hope, all ye who enter here*, for she soon made clear that the lady motorist, even with a veil, could not avoid losing the 'soft, peach-like bloom' of her complexion:

> The fact that women should motor – if such a verb may be employed – is a great tribute to their lack of personal vanity. For, try as hard as they can, it is almost impossible to make the dress they have to wear a becoming one.
>
> In the case of motor driving, there are only two things to be considered: how a woman can keep herself warm in winter and not be suffocated by dust in summer without making herself look very unattractive.

Lady Jeune continued by saying that fur was not recommended as your outer covering since it gathered dust in a manner which was almost impossible to shift. But the longer the coat was, the better, since it protected those parts of you most likely to be susceptible – namely, your extremities. Manufacturers were already seeking to solve the problem with:

> The Montague coat, a combined motoring and travelling wrap for ladies and gentlemen, offering perfect protection in all weathers, being made in specially wind and rain-resisting cloths and lined with detachable leather, fur or fleece.

Veils came in two strengths, winter and summer, and readers were advised that grey was the best colour for concealing dust. Each veil should be two yards long and three-quarters-of-a-yard wide, and should be worn with goggles to protect the eyes and give comfort. Only in her last paragraph did Lady Jeune give even a grudging acknowledgement of the pleasures of motoring:

> Those who fear any detriment to their good looks had best content themselves with a quiet drive in the park, leaving to the more ardent motorist the enchanting sensation of flying about our lovely country.

On one early long-distance trip a driver recalls: 'For our personal comfort I carried three heavy waterproof rugs, three greatcoats, thick mufflers, fingerless gloves and an umbrella' – none of which prevented them from being soaked to the skin within hours of beginning their journey.

Motoring was at first an elite activity and, as early as 1901, Harrods' General List included (in its saddlery section) outfits for motorists. Among the optional extras were brass buttons for your chauffeur's uniform, personalised with the family crest and motto. Fashion houses began taking a much greater interest in motoring fashion from about 1904 onwards, when many society beauties began to be seen in motor cars and a concentrated effort began to be made to move motoring dress from the serviceable to the fashionable. This was no doubt aided considerably by the founding of *Car Illustrated* by Lord Montagu of Beaulieu in 1902. It appears to have been a celebrity magazine of its day, which included much detail on what titled and famous motorists wore whilst motoring. This in turn had its consequences for the aerodynamics of the luxury cars of the day, where weather protection was rather better. Female fashion icons of the day tended to wear hats that were elaborate confections of feathers and false fruit, while tall silk hats were de rigeur for their male escorts. The kinds of vehicle in which such attire could be worn were anything but streamlined. Scott-Moncrieff 'recall[s] one grande dame saying, "Young man, as far as I am concerned, a sports car is any car you cannot walk through with your hat on".'

Max Pemberton, writing in 1907, offered the following guidance on motoring dress:

> A word concerning our clothes. Let us always remember that we shall rarely be too warm upon a motor, and may often be cold. Heavy overcoats should always be carried in the car, which, of course, will be provided with a hood and a glass screen. It is wisdom to dress even in summer as though spring or autumnal weather must be looked for. I believe greatly in good frieze suits for men and woollen underclothing. Wear a knitted waistcoat if you like – you will have a dinner suit in your bag, and be comfortable o' nights. But never dress lightly because the sun is shining, nor forget your heavy coats because it is not raining. These will be of stout cloth lined with fur – I do not believe in leather clothing, which is neither warm nor durable. Good cloth is at all times worth a dozen of it; while for a woman fur appears to be the only thing for heavy cloaks. True it is that in the summer we prefer the ample waterproof to the fur coat, and this is best got at Burberry's – experts in the matter, and knowing even better than we ourselves what is good for us.
>
> (Max Pemberton, *The Amateur Motorist*, Hutchinson, 1907)

The American lady motorist was an altogether more robust (or at least less fashion-conscious) breed, if the account of a Mrs A. Cuneo of New York City (in the April 1910 edition of the American magazine *Outing*) is anything to go by. Her fashion advice for fellow motorists went as follows:

> I never wear anything more than an ordinary skirt, shirtwaist and hat in warm weather, or perhaps a duster, cap, and goggles on tour. Add the necessary coat and wraps in winter, and you have all the special costuming any woman needs.

It is also worth recording Mrs Cuneo's formidable – not to say ferocious – record as a driver. After only a week's practice in her new steam tourer she entered it in the challenging Glidden Run between the White Mountains and New York. Despite crashing it off a bridge and landing upside down in a stream, she managed to complete the course. She later discovered petrol and competed against many of the leading male racing drivers of the day, beating most of them until the American Automobile Association barred women from competing in races in 1909. Among her other claims to fame were the effects she had on some of her celebrity passengers, terrifying the famous singer Caruso into silence and causing the normally fearless racing driver Barney Oldfield to scream 'Slow down!' as he was being catapulted around a wet track at Poughkeepsie, New York, during a practice run.

Fashion was even an issue for the most competitive motorist in the days before flameproof overalls became standard. Madame Camille Du Gast drove in a number of the early inter-city races. For the Paris–Berlin race of 1901 it was reported that she wore a dark leather jacket with white facings and a fur cap. Her husband travelled with her as her mechanic, but his choice of attire was not relayed to posterity. Anther-Dorothy Levitt provided one of the best combinations of fashion-consciousness and competitive spirit. She was known in society as the 'bachelor girl', famed for her luncheon parties, but this did not prevent her being a works driver for the Napier team, holding the women's land speed record (96mph in 1906) and being a champion power-boat racer. That is not to say that the idea of women racing was universally tolerated. The Brooklands rules firmly stated that 'every driver or mechanic taking part in any race shall be a male person, unless the race proposition expressly states otherwise.'

In at least one case the wrong fashion choice may have proved fatal. Count Eliot Zborowski was not only a leading early motor racer but was also noted as something of a snappy dresser. In the spring of 1903 he entered the hill climb at La Turbie and drove wearing the then-fashionable long, stiffly starched shirt cuffs. According to one account one of his cufflinks caught on the hand throttle as he was taking a bend, knocking it wide open. It caused him to crash fatally and the Minister of the Interior to panic and cancel the rest of that year's Nice Motor Week. It later transpired that Zborowski had died without having paid for his new Mercedes, but the Canstatt factory did at least have the decency to send his widow a letter of condolence and the bill for the car in separate envelopes. Zborowski's son Louis nonetheless took to motor racing, and was himself killed in a Mercedes twenty-two years later, while wearing his father's cufflinks.

In addition to its contribution to fashion, motoring has added much to the English vocabulary. We saw earlier how the French word for a stoker, *chauffeur*, came to be applied in English to someone who drove a motor vehicle for the owner. In a similar vein, the term *garage* was a French one whose introduction was widely opposed by both British and German interests. The British wanted it to be called a *motor stable* or *motor house*, while the Germans favoured the mellifluous *Kraftwageneinstellraum* (or *power wagon standing-in place*). It is hard to understand why this catchy alternative was not seized upon by the world's public.

PUBLIC ATTITUDES TOWARDS CARS AND MOTORISTS

> I use the word 'monster' advisedly, since the glaring colours in which most of them are painted, the excruciating noise they make when going at their average rate (outside Paris)

Not even canine passengers could escape from the need to wrap up against the elements. This fashion crime dates from 1903.

of at least 25mph, and the asphyxiating odour emitted by their engines, to say nothing of the appearance of their inmates enveloped by gowns and protected by hideous masks, are calculated to produce the impression of a diabolical phenomenon.

(Sir Edmund Monson, *The Times*, 31 August 1901)

Lord Montagu of Beaulieu gave us one of the earliest accounts of the public's reaction to the motor car, based on a trip he made from London to Windsor in a Panhard in 1897:

The people by the roadside were not only interested but alarmed...some people rushed into their homes, most horses cocked up their ears and in some cases attempted to bolt. Public attention, in fact, was divided between curiosity and cursing...Hotel keepers generally regarded us as people not to be admitted...One irate proprietor said he was not going to have any more of those contraptions near his place, for they might blow up at any time.

(Troubridge and Marsall, pages 81-84)

The ambivalence of the late-Victorian and Edwardian public to motor vehicles is well illustrated by an incident that occurred in the Devon towns of Lynton and Lynmouth. In May 1903 the wealthy publisher and local resident Sir George Newnes introduced a motor bus service between Ilfracombe and Blackmoor Gate, at the far end of the railway line from Lynton. It was the first time in England that a motor bus had been used to provide a feeder service to a railway, and the idea was to encourage more of the town's tourists to complete their trip by train rather than do the whole journey by horse and carriage. But within weeks of its opening Sir George found one of his drivers being prosecuted for doing more than 8mph. Sir George was infuriated by what he saw as persecution, since private motorists regularly exceeded the speed limit with impunity. He immediately withdrew the bus service and sold the buses (presumably leaving the poor driver not only fined, but also unemployed).

But opposition did not end there. The numbers of car-borne tourists visiting the towns steadily increased. Many of them had difficulty coping with the steep hills in the area, and brake failures and the resultant crashes were increasingly frequent. Countisbury Hill, in particular, earned the dubious distinction for being the most dangerous hill in the country. In 1903 local residents launched a campaign to ban cars from using it. As one resident wrote to the local paper: 'Motor cars, like traction engines and railway engines, are all very well in their proper place, but that place is not on our narrow roads.' (Quoted in Travis)

Although the campaign was ostensibly fought on road safety grounds, there was more than a hint that it was underlain by concerns that the additional volume of tourists that cars would bring threatened the peaceful – not to say sleepy – way of life enjoyed by the locals. It was not at all uncommon in those days for countrymen to complain of being suffocated by the dust from passing motorists, for their wives to complain about being unable to put out washing in gardens fronting a road and for neither of them to be able to enjoy the peace of their gardens.

At first Barnstaple Rural District Council acceded to the campaign and put forward a proposal to ban motorised traffic from the hill. But a continuation of the sleepy way of life was the last thing the town's tourism interests wanted, and, faced by the wrath of the local business community, the Council was forced to back down. A few years later they by-passed Countisbury Hill with a new and safer road.

The opposition to motor cars was not restricted to the rural backwaters of Devon. In some respects it seems that certain members of the motoring classes were their own worst enemy in the battle to win the hearts and minds of the general public. This correspondent to *The Times* makes this point:

> It is, however, a fact that in many parts of the country the motor car is still an object of hatred…It is not surprising, because there are a certain number of motor car drivers who are a curse to the neighbourhoods in which they drive. Just as there are bicyclists who 'scorch' through villages and practically deny to the foot passenger his equal right to the road, so there are drivers who seem, when they mount their cars, to put from them altogether the instincts of gentlemen…they are an exceedingly small minority of the whole body of drivers, and their display of bad manners would be unimportant, did it not tend to confirm existing prejudices and to endanger the liberty of the well-behaved.
>
> (*The Times*, 15 December 1900)

According to a correspondent to *The Times* in September 1907, proper gentlemen knew how to drive their motor cars with consideration for others. The problem was that too many cars were being purchased by those who were positively not gentlemen – 'new money', a lot of them, no doubt. Any uncertainty on that score was no doubt dispelled from some peoples' minds when it became known that Bob Crompton of Blackburn Rovers and England had become the first of a new breed of professional footballers to buy a motor car.

The House of Commons certainly contained a number of rabidly anti-motoring members. One of them told the House that the Mors make of car were so-called because *mors* meant death, and the chauffeurs of these vehicles 'believed they had a right to kill anyone or anything that got in their way.' (The coincidence that the surname of the proprietor of the company was a Monsieur Mors did not appear to have registered with him.) The splendidly named Sir Brampton Gurden, Member for Norfolk North, called for errant motorists to be flogged.

In America the public did not wait for corporal punishment for motorists to be enacted by law. The first car accident involving a child in New York in 1901 was followed by five years of anti-motorist demonstrations. A total of thirty-four incidents were recorded, including shootings, stonings and riots. Likewise, in Britain tacks or broken glass were scattered on roads to trap them, and strings or barbed wire hung across the road. There were cases of motorists being stoned, shot at and horsewhipped. In Italy, W.K. Vanderbilt nearly got lynched after running over a child, and was only saved by the use of the revolver he carried and the intervention of the local police.

The dust issue was one that affected motorist and non-motorist alike. Drivers travelled wearing combinations of helmets, goggles and veils that earned them the nickname of 'masked highwaymen' in the pages of *The Times*. They would race each other, often to the peril of other road users, to get the clean air in front and let the other driver eat their dust.

Horn sounding was a related bone of contention between motorists and non-motorists. A magistrate in 1905 ruled against the right of the motorist to sound a horn to get others out of his way, and *The Times* campaigned against the practice in 1907, arguing that: 'It is the business of everybody not to run into what is in front of him and it is the privilege of everybody not to have to concern himself about what is behind him.' (Turner, page 155) Horns were nonetheless advertised on their ability to clear the road (up to a couple of miles ahead, one claimed). They ranged from the supposedly mellifluous Gabriel horn, which purported to have a 'clear sweet tone' which was 'a welcome herald of spring in the countryside', through to the raucously unmusical Klaxon. There were even models offering horn, bugle and whistle effects driven off the exhaust. From 1908 the cacophony was added to by horns of the electric variety. The unpopularity of all these devices was undoubtedly not helped by the practice of the more arrogant brand of motorist of pulling up outside a public house and blasting them as an instruction to the yokels to come out and give them directions. But they were as popular with the motorist as they were hated by the pedestrian, and one purveyor of horns even offered a service whereby you could telephone them and have a selection of them blown at you down the line before choosing your purchase.

The revenge of the English innkeeper on the motorist was apparently had by fleecing them, if Pemberton is to be believed:

> There is no good attempting to disguise the fact that the English innkeeper believes all motorists to be millionaires, and charges them upon such a belief. The day will come when competition will help him to change his opinion and to learn wisdom…Meanwhile he is robbing himself of much profitable custom…

This was not unique to England. The first Michelin guide to France included a list of those hotels that attempted to charge car owners for overnight garaging but did not make a comparable charge for horse-drawn carriages. Motorists were encouraged not to pay such charges. Even a bath – essential after a dusty day's motoring – was beyond the organisational powers of some not-so-genial hosts:

> …the request for a bath sometimes provokes unexpected astonishment. I have even been told that there was not water enough in the house for such a mad undertaking! And I would

The horn could become a major design feature in its own right. The one on this 1904 Peugeot was designed to look as fierce as it sounded.

strongly advise all tourists to include a rubber bath among their luggage and to be firm when their demand is for water.

(Pemberton, op. cit.)

But the new mobility of motorists did bring benefits for some innkeepers. It led to the renaissance of many country inns hitherto abandoned as lost causes because of their remoteness, and some of them even went so far as to install the bathrooms that dusty motorists needed.

Perhaps the biggest problem was the tension, discussed earlier, between motorist and those who travelled by horse-drawn means. The 1896 Act gave motorists a more or less equal right to share the road with other users, though it did require the driver of a 'light locomotive' to stop 'if required to do so by any person in charge of a horse, holding up his hand.' Notwithstanding this, many motorists felt (perhaps not unreasonably) that there was an equal obligation for both motorists and those who drove horses to come to terms with the new arrangements. Knight complained that:

> The pernicious system of leaving horses unattended outside shops or public houses should not be permitted, unless the wheels be scotched and the horse secured by a rein to a sunken ring in the pavement, which is frequently done in America.
>
> It is very annoying to have to stop one's car and shout or blow the horn to call up a carter, who doubtless may be enjoying his beer in a public house, to hold his horse while the car passes by. The writer's car was on one occasion the cause of an unattended horse with a load of hay running away, and he at once wrote to the man's employers as follows: 'Had your carter not been drinking in the public house your horse would not have run away' and he has strong reasons to believe that the carter received a severe reprimand.

(Knight, page 41)

Perhaps the most creative solution to the problem of frightened horses was Uriah Smith's invention of the non-frightening car. It was fitted with what was in effect the front half of a pantomime horse. But there was a positive side to the social aspects of motoring. For all their odd attire and anti-social reputation, chauffeurs cut quite a dash with the ladies, so much so that *The Motor* reported that: 'Good looking chauffeurs are beginning to prove a very serious social problem in England, in France and elsewhere on the continent. Elderly, homely and married men are taken on in preference.'

The chauffeur's understanding of this mysterious new technology put them in every sense in the driving seat in their dealings with their non-technical employers, and invested them with a degree of glamour among the female domestic staff – and sometimes even with the ladies of the house. In America the problem got so bad that chauffer training emphasised 'the moral side of a chauffeur's duty'.

Cars were entering the public consciousness by means other than their physical presence. Mr Toad, quoted earlier, was one of the best known of Edwardian literature's motorists (and one of its worst advertisements). Cars also featured prominently in the early cinema. The first film to feature a car-centred plot was *Runaway Match*, made by Biograph in 1903. This involved a car-borne elopement and the couple's hot pursuit by the girl's irate father. It was the first of about 500 films making significant use of the motor car that were shot in the first two decades of the twentieth century, including those of the celebrity racing driver Barney Oldfield, referred to earlier. One favourite plot, picking up on the earlier point about the chauffeurs' reputation with the ladies, was that of the aristocrat masquerading as a chauffeur to win the heart (or at least the undivided attention) of the heroine. Advertisers were also quick to see the motor car's ability to convey a host of different messages – glamour, wealth, power, independence and emancipation among them – not to mention its comic potential.

CHAPTER 11

DRIVING THE CARS

In the previous chapter we saw some of the joys and frustrations of early motoring, and what relations between motorists and the general public were like. But what were these early vehicles like to drive? Few of us will ever get the opportunity to experience driving a wide range of veteran cars, but one person who did was Kent Karslake. He bought his first car, a 1912 Hispano-Suiza, in 1926. Over the years he drove or travelled in vehicles ranging from the very earliest horseless carriages to some of the most sophisticated touring and racing cars of the immediate pre-First World War period. In his 1956 book *From Veteran to Vintage* Karslake shared some of these experiences, giving a vivid picture of how it must have been to be one of the pioneer motorists. In the following sections I have drawn upon (and where possible, elaborated on) some of his accounts of driving the cars which were landmarks in the history of motoring before 1914.

1895 4HP LUTZMANN

This was a car closely modelled on the original Benz design, with a single-cylinder horizontal engine of 2,540cc producing 4hp at 250rpm (probably built under licence from Benz). The ignition was electric, it was two-speed belt-driven and had solid rubber tyres. Karslake travelled with the owner on the 1951 London–Brighton run. Before they started Karslake was given an opportunity to inspect the engine:

> Mr Berry first opened the two swing doors at the back of the boot and then propped up the lid. The single horizontal cylinder is arranged with its head pointing forwards, and under the seat, so that this extremity of the engine, usually the first to meet the eye in later motor cars, is scarcely visible. One is confronted instead with items of machinery usually hidden from the eyes of the vulgar, in the shape of the open end of the cylinder from which emerges a connecting-rod attached to a crankshaft revolving in the open air... Apart from accessories to be attended to by an oil-can, the only part of the engine to use liquid lubricant is the piston, which is fed with oil from a drip lubricator...

The carburettor was simply a small reservoir of petrol through which the exhaust pipe passed, the exhaust heat helping to vaporise the petrol. The engine simply sucks the fuel-rich air in from the top of this reservoir. Surprisingly, the machine was able to start from cold with this primitive arrangement, even before the heat of the exhaust had begun to vaporise the petrol. The owner demonstrated:

> Having set the throttle and mixture controls, and shifted the exhaust cam to the half-compression position, he grasped the rim of the great flywheel and pulled the engine round several times in order to fill the combustion chamber with mixture – if any. Then he switched on, pulled the flywheel round again until the piston was embarking on the compression stroke, took a firm grip on the rim and gave it a determined heave. The engine promptly started to run backwards, but Mr Berry stands no nonsense even from the Lutzmann, and, to my open-mouthed wonder and admiration, soon called a halt to that caper by seizing the spinning flywheel and stopping it. At the next attempt the piston successfully passed top dead centre…and the engine started to revolve in the desired direction, whereupon the intrepid operator promptly thrust his arm into the revolving machinery and put the cam over onto full compression. In a moment the engine gathered speed and was soon ticking over without a falter at a full 200rpm.

The run to Brighton was done in the rain, but they were unable to erect the leather hood because the wind was against them and it would have slowed the car's already modest progress to an unacceptable degree. None of the many old cars Karslake had driven came as close as the Lutzmann in its resemblance to a horseless carriage:

> Looking forward from the high seat of the barouche one can hardly credit the absence of shafts and quadruped. The engine and the running gear, including the side chains, are astonishingly silent, but the impulses from the large single cylinder, especially when it is pulling hard… impart to the vehicle precisely the same motion as does a trotting horse, and the illusion is intensified by the action of the springs which provide the passengers with exactly what was described as 'carriage exercise'.

Meanwhile, the driver of the vehicle had to be, as its owner put it, 'an octopus':

> Early German designers, for some reason, were inclined to completely discount the use of the feet in driving, and on the Lutzmann their sole function is to operate the push of the bell by which audible warning may be given of approach. Everything else has to be done with the hands…
>
> His right hand must be constantly concerned with the steering, performed by means of a small upright handle on the end of a horizontal lever…This steering lever is mounted on the top of a pillar, on the side of which is a horizontal lever by means of which either belt can be slid onto its fast pulley, or, with the lever in its central position, both held on the loose ones. This lever could be very conveniently operated by the left hand, were it not for the fact that the left hand has continuously to attend to the mixture control, the throttle, the advance and retard, and the hand brake.

The machine pre-dated the 1896 Act, and was therefore originally fitted with only one set of brakes – band brakes on the back wheels, which Karslake describes as 'surprisingly effective'. However, at some early stage in its life it ran away down a hill while its then owner was demonstrating it to a Mayor and Mayoress. In the ensuing accident the civic dignitaries were thrown into the road and the car's front axle was ripped off. In the course of rebuilding it the owner bought it into compliance with the law of the day by fitting it with a second braking system – pedal-operated spoon brakes on the back tyres – which were completely ineffective.

Despite being rained upon for their entire six-hour journey to Brighton, the electric ignition did not falter. Occasionally a bump in the road caused some neat petrol to slop into the induction pipe, which upset the mixture for a while, the belts slipped occasionally, but not for long periods, and Karslake only had to dismount three times to lighten the load up hills, never once having to push. On steep hills the engine slowed so much that he could count its individual beats, each of which sounded as if it might be the engine's last. He calculated that it dropped to about 120rpm. Downhill the engine could not be used to slow the vehicle as we do today, so the driver was entirely dependent upon the brakes.

Despite the absence of breakdowns, there were routine stops that had to be made – every half hour or so the fuel in the carburettor ran out and had to be re-filled by turning a tap in the engine compartment. These stops also provided an opportunity to top up the cooling water and lubricate some of the moving parts.

1897 4HP PANHARD ET LEVASSOR

Although only two years separates this car from the Lutzmann, it is a much closer ancestor to the modern car, with a two-cylinder vertical engine of 1,206cc mounted in the front. The Panhard engine (a Daimler Phoenix, built under licence) was also much higher-revving, reaching the dizzy heights of 800rpm, roughly the tick-over speed of a modern engine. The car Karslake drove was similar to the famous *No. 5* in which Levassor made his epic journey from Paris to Bordeaux and back in 1895. Over the years it had undergone some modernisation, with magneto ignition replacing the original hot tube system and the tiller steering giving way to a steering wheel. Also, for safety reasons, the original float-feed carburettor had to be replaced by a more modern one which allowed the engine speed to be controlled by a throttle. The original carburettor would have involved the engine having to be governed, which would have meant running something of that age at a dangerously high speed. Karslake regretted this last departure from the original specification, but added: 'It is reassuring to note that in spite of it the engine remains almost completely inflexible.'

The car was started by means of a handle in the front, with somewhat startling results:

As soon as it is started up…the unbalanced vertical twin imparts a vibration to the chassis which…profoundly shocked the early English technical observers who proceeded to Paris in the nineties to learn what it was the Frenchmen were up to over these horseless carriages. With its solid tyres, the car is mounted on exceptionally flexible springs…which provide very reasonable comfort even on a rough road, and the trepidations of the motor while it is running light causes the whole car to wobble like a jelly. Once the engine is under load, however, the vibration is practically unnoticeable, and was not really a feature of the car which need have caused any more alarm and despondency than it evidently did to M. Levassor.

Once installed in the driving seat, the driver had his or her hands – and this time, also their feet – fully occupied:

Seated behind the vertical steering column the drive is confronted with what at first appears to be a rather formidable array of controls. On the floor are two piano-type pedals, both of which withdraw the clutch and one of which operates in addition a band brake on the transmission. On his right, beside the seat, is a long lever which, when pushed forward, first applies contracting band brakes on drums on the rear hubs and then withdraws the clutch. Lower down is a crank handle which, by means of a screw-thread, applies spoon brakes to the rear tyres…Further forward, still on the right-hand side of the car, is another lever, which can be pushed forward or drawn back to give forward or reverse motion, with a neutral notch in the central position. The change-speed lever, between it and the brake lever, pivots on an entirely different axis.

Driving the Panhard does not appear to have been an unalloyed pleasure:

> The clutch, like most of its type and age, drags abominably, and it is quite impossible to engage the forward gear without shrieks and groans of protest from the pinions.
>
> In order to go faster one must engage the second speed, but with an engine which is most reluctant to slow down, even when its anachronistic throttle is shut, it is hard to do this silently. Exploiting the advantages conferred on us by modernity, we assisted the process by switching off the ignition while making upward changes, but of course with tube ignition this manoeuvre was denied to the driver of 1897. Perhaps in this respect at least things were easier when the assistance of the governor was available; perhaps, and I incline towards this as the more probable view, our less timid predecessors pushed the lever harder and regarded the resultant noise as just another feature of motoring…

As Levassor himself famously said of their car's gearbox: 'It's brutal, but it works…' But the steering on this vehicle was no better:

> Even at 10mph…the steering demanded all one's attention, for in spite of a wheelbase of about 6ft, which is very much greater than that of the earliest Panhards, the car was utterly lacking in directional stability. Why this should be so is not immediately apparent, for the layout of the high-geared steering…is exactly similar to that of later Panhards whose steering is impeccable…

We do know, however, that this was not the only early car that suffered from this vice. As we saw earlier in the book, the Peugeot driven by Edouard Michelin in the 1895 Paris–Bordeaux–Paris race was known as *Lightning* because of its habit of travelling in a series of zigzags.

1896 3HP LEON BOLLEE VOITURETTE

These vehicles have an important place in the early history of motoring, not only as the winning vehicles in the 1896 London–Brighton run, but also for their achievements in the Paris–Marseilles–Paris race of the same year (a race which at one stage they led) and the 1898 Bordeaux–Biarritz (which they won). More like three-wheeled motorcycles than cars, they had a single-cylinder engine and tube ignition. The Hon. C.S. Rolls had one and hated it, calling it the most troublesome machine he had ever had anything to do with. Many of his problems seemed to stem from the tube ignition, of which he said if you wanted to start out at ten in the morning you had to be up at six. It was essential that the tube should be bright red, but, in Rolls' experience, you either could not get it to light at all, could not get it to stay alight, or, at the other extreme, it lit up with a vengeance, requiring you to call the fire brigade. Once burning, according to Rolls, it made a 'noise like a cross-Channel steamer blowing off' (or, as Karslake probably rather more reasonably put it, 'a subdued but reassuring roar'). Rolls complained equally about the difficulty of starting it from cold, which he said involved cranking it for three-and-a-half hours with, at best, the handle flying off and hitting you on the nose and, at worst, a backfire putting you in hospital for two days.

St John Nixon was also less than complimentary in his recollections of the beast. The Bollee voiturette had a 640cc 3hp horizontal air-cooled engine that revved to about 750rpm and was capable of over 40mph in the hands of a brave or foolish man. Some versions of it had an engine on either side of the rear wheel, the driving of which may have required bravery tinged with suicidal tendencies. Smooth running was not a hallmark of this model. According to Nixon, 'when the engine was started, the whole machine shook and vibrated as though it were stricken with the palsy' (page 59), while Scott-Moncrieff describes the racing model as having 'a peculiar exhaust note like a quick-firing machine gun, which was both loud and penetrating' (page 40).

It was a two-seater with the driver sitting behind the passenger – who was in a particularly vulnerable position. As we saw, one of the Bollees on the 1896 London–Brighton run narrowly avoided crushing its passenger against the back of a horse-drawn wagonette, only to catapult him unceremoniously into a ditch full of muddy water.

Nixon describes some of the other unusual engineering features of this primitive machine. The gear-change was carried out by pulling a lever that moved the rear wheel forward, loosening the drive-belt. This enabled the gear to be changed, after which the lever was pushed back, moving the wheel back, tightening the drive-belt and resuming the drive. The tube ignition also meant that the engine speed could not be controlled with a throttle. This was done instead by means of a centrifugal governor operating the exhaust valve. When the engine reached a pre-set speed the flyweights of governor flew out, activating levers that prevented the exhaust valve from opening. Another notable feature of this model was that the tube ignition's burner was located dangerously near to the carburettor, which frequently caught fire, making extinguishing flames a familiar part of driving the vehicle. These components were also exposed to all the mud and grit thrown up by the rear wheel, which cannot have helped their reliability. The combination of air cooling, an un-finned cylinder head and tube ignition, according to Nixon, 'provided an almost perfect combination to bring about over-heating.' The steering was equally problematic, magnifying every bump encountered in the road, as did the unsprung rear wheel. Despite all these shortcomings, the Leon Bollee Voiturette was very popular in France.

Karslake in his trip in one experienced none of the problems described by Rolls. As for cold starting, the owner of the example he tried found it much easier with such a light vehicle to push start it.

The Bollee had a lively performance which was, according to Karslake, 'quite foreign to the average nineteenth-century motor car', but its owner acknowledged at least some of the problems and quirks identified by others:

> …for the purpose of controlling a Leon Bollee, previous motoring experience is of very little value. The art of driving, in fact, has to be mastered afresh… While his right hand is in charge of a small steering wheel, with a vertical knob on it, the driver's left grasps the spade handle of a long lever which when pushed forwards shifts the driving wheel back to tighten the belt and when pulled back to the vertical position allows the belt to run loose. At the same time, rotation of the lever successively keys each of three pinions in the gearbox to give the three forward speeds. There is no neutral position in the gearbox, and no clutch between it and the engine, so that, before the car is brought to rest, a change down must be made into the first speed, before the belt is made to run slack. Gear changing is facilitated by the fact that the lever, when pulled back, operates a brake on the belt pulley while, by means of a pedal, the driver can apply a brake to the flywheel.

Karslake also found it less noisy and more comfortable than some accounts might have you believe, but this may have been the result of nothing more than a proud owner driving his venerable machine well within its limits. What is clear is that it was a technological blind alley, incapable of being developed much further.

1901 4.5HP DE DION BOUTON VOITURETTE

These little cars were hugely popular in their own right, but were also noted for their revolutionary high-speed De Dion Bouton engine which could produce up to 6hp from a 700cc version of the single-cylinder engine, and ran at an unprecedented 1,500rpm. In some respects their design was outmoded, being rear-engined and with a four-seater *vis-à-vis* seating arrangement. Karslake naturally asked how the driver was supposed to see where he was going over the head of his passenger. The answer, apparently, was that you cannot!

(Right-hand side, three down.) The De Dion voiturette was manufactured under licence in the United States as the Pierce Motorette.

The car had an unusual layout of controls:

In front of the driver is a vertical column with, on the top of it a horizontal steering lever, mounted approximately at its centre, with at one end a vertical and at the other a horizontal hand-grip – a rather curious arrangement since both hands are seldom available for steering, and never, in my experience, necessary. Below the steering bar and on the left-hand side of the column, is another horizontal lever which, pulled back towards the driver, engages the low speed, and pushed forward away from him picks up the high. Below this again and at the front of the column, is a small lever operating the advance and retard mechanism, opposite at the back of the column is a similar lever which prevents the exhaust valve closing and thus effectively slows the engine. The only footwork required of the driver is concentrated on a pedal in the middle of the floor which also lifts the exhaust valve and then applies the differential brake. The handle of the clutch lever, if pressed vertically downwards while in the neutral position, applies the side brakes.

In some respects, Karslake suggests, having a rear engine under the seat was not such a retrograde step, giving as it did a remarkably quiet and vibration-free ride. But perhaps the most characteristic feature of this car is its steering. Karslake was warned that it was rather light:

…but in spite of this timely advice I started off, before I had got accustomed to it and to the tendency towards over-employment of the driver's hands, in a series of graceful, if disgraceful, swerves. Once one is used to it, one finds that the car steers with considerable precision if one remembers to be really light-fingered with the lever, but the steering resembles that of some more modern cars in that there is curiously little 'feel' through it of what the car is doing. As a matter of fact, with a wheelbase of 5ft 1.5in and 700x80mm tyres blown up pretty hard, it seems to me rather surprising that the car steers at all, considering what an enormous proportion of its weight must be on the back axle.

This minor vice aside, Karslake found the experience of driving the voiturette 'delightful'. Whilst it was by no means a high performer, particularly on hills: '…for unhurried use in fairly flat country it would almost certainly prove an entirely satisfactory vehicle, capable of keeping going day after day with the minimum of trouble.'

Nixon recalls some models of the De Dion Voiturette having particular difficulty starting from cold:

> This was due firstly to the very poor quality of the early coils and secondly, and mainly, by the fact that the carburettor was separated from the engine by a very long inlet pipe which was entirely unheated. It is an accepted axiom among modern designers that the carburettor shall be as near the engine as possible, but the inlet pipe on these De Dion Voiturettes was probably a couple of feet long, which made starting very difficult, particularly in the cold weather.
>
> (Nixon, page 57)

They were also prone to clutch-wear, which could be serviced, but only, on the early models, by taking the whole gearbox out.

1904 12HP LANCHESTER

As we have seen, Dr Frederick Lanchester is a name held in high renown among early car designers as the designer of the first real British car. And yet, while his cars anticipated features that would not appear on other makes for up to fifty years, in some respects he clung to aspects of outdated technology long after other makers had rejected them for better alternatives. A good example of this is tiller steering; something that he did not abandon until 1911, albeit a markedly developed version. Karslake drove an early 1904 model with a 4.1-litre two-cylinder engine producing 12hp at a leisurely 760rpm (later 1904 models had a four-cylinder 2,470cc engine). He struggled at first with the unfamiliar controls:

> In the first place, steering is by tiller, but unlike those on most early cars where the fulcrum is in front of the driver and the lever is turned away from the direction in which it is intended that the car should proceed, on the Lanchester the tiller springs from behind the driver and on his right. His right arm in fact lies along it, and he pushes it the way he wants the car to go, which, in the words of a contemporary authority 'makes oversteering practically impossible owing to the centrifugal force which acts on the driver's body and tends to counteract his steering effort'…all that is necessary is to look in the direction where one wants to go, whereupon the car goes there, much as a bicycle does when one is riding it.

The right hand is thus fully occupied, but there is little for the driver's feet to do. The left foot operates a control (when needed) that cuts out the action of the engine governor, while the right one does nothing more than stamp on the bulb of the horn. The real hard work is reserved for the left hand operating two levers which by a laborious process first pre-select and then engage the different gears. In between doing this, the left hand also adjusts the mixture control, to which the engine is particularly sensitive, and operates a plunger pump which maintains the fuel supply to the wick carburettor (another of Lanchester's more esoteric features). At particularly busy times it is suggested that the assistance of the front seat passenger is enlisted for some of these tasks. But some of those who had Lanchesters could not speak highly enough about them. Here is one happy owner, himself an engineer, quoted by Pomeroy:

> The special advantages of the Lanchester are silence in running, great comfort on account of the method of spring suspension, absence of chains, absence of vibration from an imperfectly balanced engine, simplicity of driving and consequently but little fatigue, safety owing to a broad wheel gauge and low centre of gravity, its perfect brakes, and an ignition system that does not develop faults…. I have never seen any piece of complicated mechanism… better designed or with better material than the Lanchester.

> (Hough, page 204)

1904 18–28HP MERCEDES AND 1903 60HP MERCEDES

It was the 1901 Mercedes that was said to have rendered every other car then in production obsolete. In particular it was their 60hp production car that was enlisted after their entire team of 90hp racers were destroyed by fire shortly before the 1903 Gordon Bennett Trophy. Stripped of their bodywork, but otherwise little modified, these 9.2-litre four-cylinder cars proved good enough to beat the best race-prepared machinery that France, England and America could throw at them. Karslake had the opportunity to drive both a Sixty and its smaller counterpart, the 4-litre 18–28hp model, the particular example of which had itself been raced with some success at Brooklands.

The 18–28hp was fitted with a scroll clutch, a speciality of Mercedes:

> This is sometimes a little disinclined to free completely, so that a gear has somewhat to be prodded in while the car is stationary, but it gives a smooth take-off which is excellent by any standard and which must have been quite delightful when compared with the usual fierce cone clutches of the period. The only objection to it is that if the throttle is opened at all violently as soon as the clutch appears to be home, it is prone to slip, so that a racing start is impossible and there is some delay after every gear-change. This is more noticeable because the gear-change itself is absolutely delightful.
> …the engine is delightfully flexible, but the top gear is decidedly high and for best performance one needs to make ample use of the gearbox. One is moreover encouraged to do so not only by the ease of the gear-change but also by the quite remarkable willingness of the engine to rev. (up to 1,500rpm).

The great disappointment was the steering. The 1904 model, unlike that of the years before and after, was fitted with three-quarter elliptical front springs, possibly to give a better ride over bad roads. However, these did nothing to locate the front axle and, if a front wheel hit a bump while cornering, it completely threw the steering out, causing the car to yaw alarmingly. This is, however, readily corrected, and is more disconcerting than dangerous.

But it was the Sixty itself for which Karslake reserved his most enthusiastic praise. It displayed all the virtues of the 18–28hp, and its steering, whilst heavy, was at least accurate. But there was more – much more:

> Where the 'Sixty' really demonstrates its peculiar quality is in the almost unbelievable low-speed torque developed by its mighty engine. Here is no stream of power such as is let loose by the best six-cylinders of a later date, but, when the throttle is opened, a series of giant impulses, smoothly delivered but each individually appreciable, giving an impression of irresistible

power and communicating an intoxicating sense of omnipotence to the driver. Now one can understand why the 'Sixty', although good drivers declared it to be the safest car that at that date had ever been built, was the death of so many less experienced drivers. It is not good for the uninitiated to receive the impression that they walk – or motor – with the gods…

1904 GARDNER-SERPOLLET STEAM CAR

Leon Serpollet built his first steam car in 1887 and three years later completed a 286-mile journey from Paris to Lyon, albeit in five days. F.L. Gardner was the American financier who enabled him to put this vehicle into production, though he had previously produced a steam car in collaboration with Armand Peugeot for the 1889 Paris World Fair. The 1904 model was indistinguishable on the outside from an internal combustion-engined car of that period. The flash boiler on the back looks as if it could have been a tool box. Only when you came to light the boiler did the differences become clear:

> There are several ways of doing this, from squirting methylated spirit into the boiler box and putting a match to it, to applying a gas poker to the situation until the paraffin becomes sufficiently volatile to be sucked up through the burners and get going in the proper furnace style. In either case one is confronted with naked flames of a character calculated to terrify the motorist accustomed only to internal combustion, and, indeed, he must learn at times…to see these flames leaping up the back of the body nearly to the height of an elegant Edwardian leather hood without registering undue alarm…
>
> Once steam is up, however, it is impossible not to fall for the essential refinement of a steam engine…By merely pushing the lever which admits steam to the engine and opening the throttle one is assured that the car will glide away from rest with the utmost silence and smoothness. A Silver Cloud Rolls-Royce may compete with it in this respect upon absolutely equal terms, but in the context of 1904 the performance is little less than uncanny.

The image of Serpollet's steam cars was no doubt helped by the fact that one of them became the first non-electric car to hold the world land speed record. In April 1902 his car *The Easter Egg* set a world speed record of 75.06mph in rather an unlikely setting for such an attempt – the Promenade des Anglais in Nice, France.

The 1904 Gardner-Serpollet steam car.

Serpollet's steam cars found favour in high places. Both the Shah of Persia and King Edward VII bought one. The problem (though not one that would have worried the aforementioned purchasers) was that they took a great deal of skill to drive. The throttle had to be coordinated with the operation of a donkey engine that sent more water to the boiler to make more steam, and the need for steam – for increased speeds or for climbing hills – had to be planned at least a little way ahead. Too little steam and you were short of power; too much and it blew off wastefully through the safety valve. This and the amount of maintenance they required were major factors in ensuring that they did not become a major force in modern motoring. The rate of their evolution could not match that of the internal combustion engine. Despite this, Serpollet continued to champion the cause of steam until his death from consumption in February 1907.

1908 GRAND PRIX ITALA

The Itala was built in Turin from 1904 onwards (Itala was an Italian concern, but also had a factory at Brooklands). They very quickly entered racing, winning the Targa Floria in 1906 and the Peking–Paris endurance race of the following year. The particular model Karslake tried is one of the giant Grand Prix racing cars that took part in the 1908 Grand Prix at Dieppe. It came in the red bodywork of the Italian team, the national racing colour that that nation took over from the Americans in 1907. Under the regulations of the time the maximum permitted bore of the four-cylinder engine was 155mm, and they opted for a stroke of 160mm, giving them a capacity of 12,076cc and 100hp at 1,800rpm. This capacity was smaller than the rival 14-litre Bayard-Clements and 13-litre Mercedes, and the Itala was also the heaviest car in the race. This may explain why the most successful member of the team finished no higher than eleventh.

The engines in the early Grand Prix cars rapidly grew to monstrous dimensions. This 1908 Panhard-Levassor, of a kind that may have raced against the Itala, had a capacity of 12,831cc.

Given that each cylinder on the car was more than 3 litres, one might expect that the use of the starting handle might be lethally dangerous or require superhuman strength. Neither is the case. The car was fitted with a sliding camshaft to reduce compression. Once this was in use, the engine could be turned quite easily, and the real drama only started once it was running: 'after a few turns the giant engine usually bursts into life with a shattering roar, so startling even when expected that it requires great presence of mind to remember to slide the camshaft back onto full compression.'

But the drama of starting it was not matched by the experience of driving it:

> …the Itala is not only one of the most exciting cars that it has been my lot to drive, but also one of the least alarming. The steering is of thoroughbred precision, with no trace of undue heaviness in spite of the fact that the wheel only needs one complete turn from lock to lock. The gear lever is pushed or pulled, rather than flicked, in its gate, as if to indicate that there is something pretty solid in the way of pinions on the other end of the mechanism, but it moves with complete smoothness, and the gears change with unfailing ease and silence for any operator versed in the first principles of sliding pinion gearboxes. The throttle pedal, it may be noted, is placed, in what was to become traditional Italian style, between the clutch and brake pedals, and, more exceptionally, appreciably nearer to the driver, in a position which permits of ideal heel and toe operation of brake and throttle whilst slowing and changing down.
>
> Yet there is one insidious peril in the driving of this car. There is, it must be remembered, 100hp available to the driver, delivered at only 1,800rpm, which means that when the engine is exerting a really tremendous urge, it sounds and feels as if it was doing next to no hard work at all. The effect of this on the driver, at least until he is accustomed to it, is an intoxicating sense of power without responsibility.

1912 MODEL T FORD

We move from the Italian racing elite to one of the first people's cars. When Ford announced his product range for 1908, he promised 'one new runabout – a light roadster that will show the way to everything else on wheels, quality, price, performance and appearance beyond compare'. The Model T Ford was originally launched in the United States in that year, and was assembled from imported components in Britain from 1911 to 1927. It sold in Britain for £175, for which you got a four-seater open-topped car with a four-cylinder 2.896cc engine delivering 20bhp at 1,600rpm. It had just two gears and (in its latter years) an archaic chassis design, but was rugged, reliable and backed up with an outstanding dealer network. It was not always the case that the car was available in 'any colour, so long as it is black'. Originally, you could also have them in grey or red; black was only introduced as the standard colour because it was the paint that dried the quickest.

Driving one was unlike handling any other car, according to those who knew them. There was no accelerator pedal, but rather a hand throttle and an ignition advance and retard on the steering column. To stop the driver pulled a lever on his right-hand side, which at the same time locked the gears in neutral. To start off again the driver had simultaneously to release the hand lever and press the left-hand pedal down to engage the lower of the two speeds:

> The lever has meanwhile pre-selected the high speed, and releasing the pedal will allow this to engage. To do so smoothly, however, requires some skill, for the gap between the two speeds is substantial, and I well remember the typical Model T driver of the past, who, after letting the engine roar away on the low speed, would suddenly take his foot off the pedal, proceeded for the next few yards in a series of sickening jerks…such interruptions to the passengers' comfort are quite unnecessary, if you have once mastered the technique of letting the car gather way on the low speed and then momentarily closing the throttle before easing the high speed in with the pedal, after which the throttle can be smoothly opened again.

Once he is safely 'on high'…the driver has really little more to do in all ordinary circumstances but to steer and operate the hand throttle…

It is when you come to stop, however, that the designer's peculiar arrangements really make themselves manifest. In order to stop, you have to tread on things – preferably on the gear (or more properly clutch) pedal sufficiently to take it to the neutral position between high and low, and on the foot brake; but Henry Ford was convinced that some of his less-experienced drivers when faced with an emergency would remember to tread but might not be too particular where they were treading. He therefore arranged matters so that it did not matter very much where they trod. The brake pedal was of course all right; but if they bore too heavily on the clutch pedal, it would only engage the low gear, which at least would slow the car, while if they chose the remaining pedal, which engages reverse, they would check forward motion even more effectively. Indeed their action would not only stop the car but immediately thereafter set it going backwards…

If all else failed, and you did not mind risking the tyres being torn, there was always the hand-brake.

It was not a car that was over-endowed with extras. Instead of a petrol gauge you were provided with a graduated ruler to dip in the tank. There were no shock absorbers, no water pump and the driver's door was merely a non-opening shape, pressed into the bodywork. The petrol relied on gravity feed rather than a pump, and the engine lacked an oil pump, relying instead on a rudimentary 'splash and spit' system of lubrication. This could mean that, if you were going up a steep hill, the engine could be both starved of fuel and in danger of seizing up, and it was therefore advisable to take hills backwards. Another idiosyncrasy was its tendency, when being started, to run over the person cranking the starting handle.

A CATALOGUE OF HORRORS

So far in this chapter, we have looked at cars that, whatever their shortcomings, played a positive part in the history and development of the motor car. The earliest days of the motor industry threw up many men of great skill and foresight who can take the credit for the motor car's rapid evolution. But it also attracted some whose ignorance of the most basic engineering principles was so profound that, in a just world, they would have been barred by law from ever touching a spanner. The vehicles they produced represented the lowest points of the early motoring experience.

One of the most entertaining and comical pieces of writing about the early days of motoring is that part of St John Nixon's book *The Antique Automobile* in which he describes, often from bitter first-hand experience around the turn of the century, some of the most dreadful vehicles ever to appear – usually only briefly – on the roads. I commend the whole chapter (and indeed the entire book) to you, if you can still find it, and the following sections can do no more than whet your appetite by introducing you to some of the grisly exhibits in his 'museum of horrors'.

Perhaps not surprisingly, Exhibit A consists of the complete works of Edward Joel Pennington (1858-1911), the confidence trickster whose involvement with Harry Lawson is recorded earlier in this book.

Most of Pennington's inventions were barking mad. He produced flying machines with the aerodynamic qualities of a brick, motorcycles that would not run for more than a mile or so, a spark plug that defied all the laws of metallurgy and scarcely survived from one lamp post to the next, and a 'powered' bicycle driven by a propeller mounted behind the driver's back. But this did not stop Pennington speaking very highly of his brain-children. His motorcycle, Pennington claimed, was able to leap 65ft through the air given the slightest incline to jump off. His first car was made out of two ladies' bicycles joined together by a platform and powered by two separate motors, one driving each of the rear wheels. It seemed that Pennington only had to be shown accepted wisdom for him to conclude the opposite. As Nixon put it, even at the time Pennington was working there were certain basic principles to which most car designers subscribed. For example:

- that cars should be steered by the front wheels and driven by the rear ones;
- that petrol engines required a carburettor of some kind; and
- that the engine's flywheel should revolve vertically, rather than horizontally.

Pennington naturally assumed the opposite. The next car he designed, the Raft Victoria, had front-wheel drive and rear wheel steering. The engine was mounted horizontally amidships and was entirely un-cooled by either air or water. On the contrary, many aspects of its design seemed to be positively aimed at retaining heat. The horizontal flywheel was made from several pieces of metal rather tenuously riveted together, which wobbled alarmingly from new and soon developed a tendency towards disintegration. In a similar vein the long, frail chain drive from the engine was so designed as to work loose very quickly and start chewing the chain wheel to pieces. Pennington thoughtfully provided a lever which enabled the driver to loosen or tighten the chain as required. However, one further piece of genius in his design was that if this chain had to be tightened then pulling the lever automatically loosened the belt that relayed the drive to the front axle. If one pulled the lever the other way to tighten the belt it correspondingly loosened the chain, requiring the driver to stop and take some links out of it.

Pennington also dispensed with the superfluous convention of a carburettor, preferring to drop the fuel by gravity through a needle valve immediately above the inlet valve. This worked after a fashion until the (un-cooled) engine was put under a heavier load (say, by going up hill), at which point 'the unfortunate sparking plug was promptly incinerated. The average life of a sparking plug in this engine was said to be not more than three miles.' (Nixon, page 135) What's worse; these were not even Pennington's own inoperable design of sparking plug.

Pennington's rear-wheel steering also had to be unconventional. In place of a steering wheel or a tiller he placed a vertical tube to the right of the driver, which had to be pulled or pushed according to the direction in which you wanted to turn the car, assuming you could remember whether pushing the lever turned you left or right. Starting was another hit-or-miss adventure, using a detachable starting handle that went through the foot-boards and often physically could not engage with the fitting on the crankshaft.

Notwithstanding these minor shortcomings, an attempt was made in 1899 to drive a Pennington car from Manchester to the Agricultural Hall in London, where one of the early motor shows was being held. The driver, H.W Egerton, whose account of the ordeal we saw earlier, took with him 'enough spare parts to make half a dozen cars on the road, including four dozen [other accounts say seventy-two] spare sparking plugs. By the time he had reached Wilnecote, near Nuneaton, he had used up everything and he was forced to "abandon ship" and come on to London by train' (Nixon, pages 135-6). This did not prevent Pennington advertising at the show that his car had come all the way from Manchester (neglecting to explain that much of the journey had been by rail).

Such were the engineering skills for which Harry Lawson handed over £100,000.

Next in the museum's 'catalogue of horrors' was the so-called Orient Express. This was built by Bergmann's Industriewerke of Gaggenau, Germany. As Nixon explains:

> It had at the rear a single cylinder, horizontal, water-cooled engine with a bore and stroke of 5 x 6 inches, which was said occasionally to produce four horses in point of power. This engine was of the 'one-explosion-per-telegraph pole' type, and each such explosion was, to the unfortunate passengers, somewhat like a kick in the pants.

(Page 138)

One from the chamber of horrors – the 1898 Orient Express 4hp, which seemed to consume more water than some steam cars.

It had an unusual system of thermo-syphon cooling which, combined with a radiator safely hidden away from the slightest hint of air movement, gave it 'a thirst which would be the envy of the most confirmed inebriate.' In a fifty-mile efficiency trial in June 1899 the vehicle consumed no less than five-and-a-half gallons of water, two gallons more than the amount of petrol it required for the journey. But the transmission was the car's *piece de resistance*. There were a total of four flat belts and eight pulleys connecting the crankshaft and the countershaft, so, to change gear, the driver had to:

- release a lever (resembling one of those found in an old-fashioned railway signal box);
- close the hand throttle slightly;
- move a pre-selector lever to the right notch;
- pull the signal box lever; and
- reset the throttle.

All of this had to be done with the hands which were also supposed to be steering. Any error in pushing the pre-selector in could result in more than one gear ratio being selected at the same time, which was not to be recommended.

Exhibit C comes from a British manufacturer – the Marshall. Built in Manchester, it was a bad copy of a French bad copy (the Hurtu) of one of the more primitive Benz cars – a promising pedigree. This shared the Orient Express' vice of engaging more than one gear at a time, but it also brought new joys to motoring. One was a particularly bad version of the surface carburettor, which had a separate control for adjusting the fuel-air mixture and which required the more or less constant attention of the driver. It was apparently particularly temperamental on bumpy roads (of which there was no shortage in Victorian and Edwardian England). Its performance also left much to be desired. On the outward leg of a fifty-mile reliability trial in 1899, it managed to average just 5.4mph (though on the return journey this was increased to a dizzying 8.2mph). It was perhaps as well that its performance was modest since it was found to require more than twenty-four yards to stop from a speed of just 6.99mph on a gradient of 1 in 12.9. As for climbing hills, the only way up was to push it or tow it. The judges on the trial found its performance 'very unsatisfactory', though this did not deter the manufacturers entering two of them in the much more challenging 1,000-mile trial in April/May 1900. The records unsurprisingly suggest that they did not complete all the stages, though one of them, for some reason, is said to have won a £10 *Daily Mail* prize. A booby prize, perhaps?

But at least it was not prone to catching fire, unlike Nixon's next exhibit, the Stirling Dogcart. We will pass rapidly over the shortcomings of the cooling system (another radiator cunningly concealed from the fresh air, with no water pump and water cooling only to the cylinder head – problems which it overcame by carrying a tank which held enough water to float the *Titanic*). The fire hazard came from having tube ignition whose burner was located right next to a carburettor, whose design faults caused it to overflow at every opportunity. The only safety feature was that the tube ignition on this model, which – in addition to its tendency to be blown out by the wind, something common to all such systems – also tended to be extinguished by bumpy roads. So, when the carburettor did overflow, there was at least more than a sporting chance that it would not catch fire.

The 1901 Stirling Victoria de luxe, younger brother of the Stirling Dogcart, with its alarming ability to catch fire.

Over and above this, the supposedly internal workings of the engine were exposed to the air and it also had the failsafe of two separate braking systems – except that they were only failsafe in that neither of them was safe and both were prone to fail. The one that acted on the countershaft would not work if one of the drive chains broke or came off; while the primitive spoon brakes that rubbed directly on the rear tyres failed if those tyres punctured or were under-inflated.

The mechanics of the car were located much too far to the rear of the car and this, combined with a clutch that was either fully in or fully out, meant that the car's front wheels tended to lift off the ground, dragster-style, as the car set off. This had serious implications for the steering: suddenly there wasn't any. There were no manual engine controls of any kind. Like most cars of the period, the speed of the engine was controlled automatically by a centrifugal governor that cut in as the engine speed reached a certain level. But unlike other models, the Stirling had no way of overriding the control, for example to get the engine to speed up for climbing a hill. This resulted in a series of jerks, as the speed of the car overtook that of the cut–out engine, causing it to cut back in until it in turn overtook that of the car and stopped again, with more jerking.

The steering of the Stirling was primitive in the extreme – the whole of the front axle pivoted on a single beam, like a child's soap-box cart, with the result that, if you hit a brick or a bad pothole, the steering could be wrenched violently out of the driver's hands, leaving you in a ditch – or worse.

Nixon's experiences of this car included having a mechanic drive it backwards from Regent Street to Clapham Common to have a new set of gears fitted after all three forward speeds failed. He was once asked to deliver one to a customer in Liverpool from the Regent Street showroom. Nixon did not view such a prospect with much hope of success: 'A non-stop run on this car from Regents Street to Golders Green would have been something of a feat; to Barnet, almost unique experience, and to St Albans a red-letter day in one's life. Liverpool was a fantasy.' (Nixon, page 146)

In the event the car broke down comprehensively within 100 yards of the showroom and all further thoughts of Liverpool were abandoned.

Nixon describes the Victoria Combination as 'not dreadful in any sense of the term…just an impossible contraption so utterly futile in its general layout and hopeless in its conception.' Its hopelessness really came into its own when the driver tried to stop it. The brakes were of the contracting variety; leather lined and operated by a wire on drums the size of small saucers. So ineffective were they that Nixon found he was able to stop more effectively by sticking his foot out of the vehicle and scraping it along the ground. But if stopping was potentially lethal, starting was also not without its dangers. The high-compression De Dion Bouton engine with which it was fitted had a starting handle about 4in long and with practically no leverage. A backfire whilst using this meant a dislocated thumb or a broken wrist, and the only safe way to get it going was a push start in top gear. These French vehicles 'were imported into this country in extremely small numbers, a blessing of which the public were entirely unconscious.'

One that just failed to make it into Nixon's museum of horrors was the MMC motor-tricycle. This was despite the fact that the example entered into the 1900 reliability trial needed so many replacement parts en route that, it was popularly rumoured, the only part of it to both start and finish the trial was a single switch.

CHAPTER 12

TECHNOLOGICAL BREAKTHROUGHS AND BACKWATERS

In this chapter we take a look at how the features of the modern car that we take for granted evolved. These early years were ones of rapid technical development and, inevitably, that development did not always go in the right direction.

ENGINE DESIGN

The period leading up to the First World War produced some of the most dramatic increases in efficiency seen in the history of the motor car. In the decade up to 1910, the power output of a typical four-cylinder engine increased by about three-fold. The earliest internal combustion engines were exceedingly heavy, slow-running and not particularly powerful. They produced only about 1hp for every 300lb, and ran at no more than 250rpm. Daimler's 1886 engine design would reach 800rpm and produced 1hp for each 90lb. Within a few years De Dion Bouton had further developed the design with the use of light alloys, to give engines that ran at 1,500rpm and produced 1hp for each 25lb. But the higher power and operating speed put far more stress on the mechanical components, and on the cooling and lubricating systems that kept them working.

Over this same period multi-cylinder engines became increasingly the norm. There were limits to the amount of power that could be squeezed from a single-cylinder engine by increasing its size, since the moving parts in it became bigger and heavier. The bigger engine therefore had to run more slowly, and its power did not increase in proportion to its size. Although De Dion Bouton made considerable strides with them, it was difficult to design one which could run at high speed for any length of time without running the risk of shaking itself to pieces. Manufacturers moved to multiple cylinders, despite the greater difficulty of manufacturing them.

Before the start of the century four-cylinder engines were largely confined to racing models. But they spread rapidly into road-going vehicles. Mors built a V4 in 1896 and in 1904 Napier had the first practicable six-cylinder engine. De Dion Bouton followed with a V8 in 1910.

It may be worth at this stage explaining the term *horsepower* and its use in describing the power of early motor cars. The idea was originally developed by James Watt for his steam engines. He borrowed some strong horses, used for pulling brewers' drays, and carried out some experiments to find out how much work they could do in a minute. This he calculated at 33,000ft-lb (the equivalent of lifting a 1lb weight through 33,000ft, or 33,000lb through 1ft, or – more realistically – some combination of the two). In fact, we know that a horse can exert much more power than that for short periods – possibly three or four mechanical horsepower – but it cannot continue to do so all day, every day, in the way an engine can.

The power of the engine was measured at the flywheel or the crankshaft, and a good proportion of it – in some experiments with early cars, up to 40 per cent – was lost between the engine and the driven wheels, as a result of mechanical inefficiency in the drive trains, with their complicated arrangements of belts, chains, gears, driven shafts, and so on.

Knight provided some simple calculations for his 1902 readers, to help them to convert their needs to horsepower. First he points out that the road surface makes a lot of difference. On a good, level, macadamised road he estimates that a car weighing one ton fully laden would require a tractive effort of 50 to 60lb to move it at 10mph. Rubber tyres would reduce the effort required somewhat, pneumatic tyres some more, whereas dust, mud and stones would greatly increase it; so he allowed an average of 70lb on a dead-level road. 10mph is 880ft per minute, so the horsepower needed to propel this vehicle at 10mph is 70 x 880 = 61,600ft-lb, or just under 2hp. But if up to 40 per cent of the engine's power is lost through the transmission then the engine would need to be more like 2.75hp. Come to a hill and the situation changes. On a 1:10 hill the car rises 1ft for every 10ft it travels forward, so one-tenth of the weight of the car – or 224lb – therefore has to be added to the calculation. Thus, to go up a 1:10 hill at 10mph requires less than 11hp once the inefficiency of the transmission has been factored in. However, Knight suggests that, given the extra strain of hill climbing on the machine, something nearer to 16 or 17hp would be appropriate. All of this explains why, on some of the earliest cars, passengers often had to get out and sometimes even push on hills.

Two-stroke engines were developed at about the same time as the four-stroke. Sir Dugald Clark invented one in 1879, and Benz himself made one. Although they were simple and worked quite well at lower speeds, they were also inefficient and dirty (in terms of emissions). Their use was largely limited to motorcycles and the smallest cars.

It took a quite considerable time for the petrol engine to establish its supremacy over its rivals. In America, in particular, the cleanness and reliability of the electric car made it particularly attractive to the lady driver. And as late as 1906, *The Motor Yearbook* could say: 'Advocates of steam cars have, on the whole, every reason to be satisfied with the way in which their prospects have improved during 1905, and with the extent to which revived interest has been taken in them by the buying public.'

It should not be thought that the hybrid car is a recent development. Over a century ago an Austrian carriage-builder called Ludwig Lohner decided to enter the electric vehicle market. He employed a young man who would become better known in a different area of motoring – Ferdinand Porsche. But the batteries increased the weight of the car excessively and affected its handling. They therefore went for a so-called mixed-car, in which a Mercedes petrol engine drove a dynamo, which in turn activated two electric motors set into the front wheels. Panhard–Levassor acquired some of the overseas patent rights for this in 1902.

Eccentricity also flourished. As we saw, Harry Lawson introduced a self-contained propulsion unit called the 'motorwheel' in 1899. It could be used for a variety of purposes, including replacing a pony between the shafts of a trap, and consisted of a single-cylinder air-cooled engine mounted in a light frame which also contained a single road-wheel. Best of all, in 1906 a Frenchman named Constantin came up with a set of roller skates, into each of which a tiny petrol engine was fitted. The petrol tank and ignition arrangements were strapped to the back of the 'driver', who could – if brave or foolish enough – reach a potential top speed of 30mph.

FUEL AND AIR

A key part of the internal combustion process was to create the right mixture of vaporised petrol and air for efficient combustion. This is not a simple process since different mixtures are needed for starting, slow running, maximum fuel economy and maximum performance. The earliest cars used a variety of methods to do so. The 1901 Delahaye bubbled air, heated by the exhaust gases to aid vaporisation, through a reservoir of petrol. The petrol-rich air then passed through a hand-operated mixing valve where it was combined with fresh air to produce an explosive mixture of the required strength. The Lanchesters of 1897 to 1914 were among those who used a system involving wicks to mix petrol and air – the Lanchester was one of the last to abandon this method. Siegfried Marcus, whose 1875 motor carriage anticipated many of the technical advances that would later appear on the 1886 Benz, used a revolving brush to agitate the petrol against a surface and help it vaporise. As we recorded earlier, the appalling Pennington car dispensed with a carburettor altogether.

But it was Wilhelm Maybach who is generally (and wrongly) held to have come up with the design of carburettor that was used on most of the twentieth century's cars – wrongly, because a court held in 1901 declared that his 1893 designs had been anticipated in most important respects by the 1887 and 1889 patents of a Briton, Edward Butler. But whoever gets the credit for it, the principle of the spray carburettor was that, during the induction stroke air was drawn into the carburettor venturi – a funnel-shaped intake in the carburettor. Meanwhile a steady fine stream of petrol was also drawn into the choke chamber from an adjacent reservoir, where it mixed with the air and atomised into a combustible mixture. A butterfly valve was then used to control the amount of this mixture going into the cylinder.

When petrol evaporates heat is absorbed from the surrounding air. This reduces the temperature of that air and makes further evaporation more difficult. If the surrounding air temperature is low enough to begin with, this can also cause water vapour or even ice to form. For this reason the Maybach carburettor, in common with many of these early systems, used hot exhaust gases to keep the operating temperature up. Almost all cars after 1900 used a Maybach-type spray carburettor or developments of it.

The early engines tended only to operate properly at one speed, which was more or less flat out. This was largely because the early carburettors could only supply a fuel-air mixture of a single strength, which was either too rich or too weak when the engine was working at different speeds or loads. Therefore, instead of a throttle, they had an automatic governor that worked by centrifugal force, limiting the speed of the engine to a pre-determined level by restricting the operation of the exhaust valve. This meant that a full charge of fuel could not be let in through the inlet valve. When the driver needed to slow down he would use a hand- or foot-operated accelerator which changed the setting at which the governor took effect. This made the engine run unevenly. This basic model of the carburettor was therefore further refined over the years. A more petrol-rich mixture was needed for starting, and the choke control was added to restrict the flow of air into the chamber and provide this. Additional jets were also provided to allow more or less fuel to be injected. These refinements gradually made the governor superfluous and enabled the driver to the control engine speed with the throttle.

BREATHING

A number of the earliest cars had atmospheric inlet valves. As the piston went down on its induction stroke it created a partial vacuum which sucked the inlet valve open and allowed the petrol-air mixture to enter the cylinder. Whilst this worked well enough on the slower early engines, on more advanced models such valves opened too late and closed too early, preventing the optimum amount of fuel and air reaching the cylinders. They also only worked efficiently within a narrow band of engine speeds, which more or less forced you to have a fixed-speed governed engine. To achieve higher engine speeds and better performance, mechanically operated valves had to come into general use.

Various systems were tried, such as the sleeve valves favoured by Daimler and the rotary valves introduced by Darracq and Itala, but the focus for development was the poppet valve, which eventually became the industry standard. At first these were located on opposite sides of the cylinder, but side valves proved simpler to operate and were the norm by 1914. Overhead valves worked better still, but the engineering for them was more difficult; they were harder to lubricate and made maintenance more tricky, as well as being noisy. Hence, only about 10 per cent of pre-1914 cars – generally the higher performance models – had them. Other supposedly 'modern' innovations, such as twin overhead camshafts and four valves per cylinder, were also to be found in some of the best pre-1914 models.

As with so many aspects of its design, the Lanchester, which appeared in prototype form in 1897 but which was not offered for sale until 1901, had valves and gears that were like nothing made before or since. There was just a single valve per cylinder, through which both inlet and exhaust gases passed. On most cars of the day the lack of the technology needed to make steel suitable for operating at the high temperatures of a car engine, meant that the valves would need to be re-ground, if not replaced entirely, every 1,000 miles or less. The Lanchester's unexpectedly efficient system would last several times as long. So, whilst the system did not point the way ahead for car design in the long term, it dealt efficiently with the technological limitations of the day.

A simple way to increase the power of an engine is to increase its compression ratio, but there were problems with this for the early manufacturers. First, it was more difficult to make a high-compression engine gas tight, which was one of the reasons why many engines had the cylinder head cast as a single piece with the block (although the compression also leaked past the pistons). Second, this also increased the amount of friction created by the piston, creating a challenge for the early lubricants. Third, the fuels of the day could not cope with a compression ratio of much more than 4:1 before they started knocking. But designers like Lanchester discovered that helping an engine breathe more easily could, as he put it, 'turn a pig of an engine into a willing horse', and the principles of good breathing were gradually worked out and incorporated into standard engine designs.

LUBRICATION

Some of the earliest models, like the 1895 Lutzmann described elsewhere in the book, had an open crankcase. The bearings and cylinders were drip-fed with oil from individual lubricators. Once the crankcases were sealed, lubrication was mainly by the 'splash' system, whereby scoops on the crankshaft dipped into a bath of oil and literally threw oil about inside the engine. Ducts were fitted inside the engine in an effort to direct more of the oil to those parts of the engine most needing it, and various ingenious methods were adopted by the manufacturers to equalise the oil flow around the engine so that, for example, it did not all run to the back of the engine when the car was going uphill. Even so, it was a very hit-and-miss system and not up to the increased lubrication demands of the higher performance engines that were beginning to appear. It was also a very high-maintenance system, requiring the oil to be changed and the system flushed out every few hundred miles. It was difficult to gauge how much oil there was in the engine and, because the big fear was that the engine would seize up from too little oil, the tendency was to use too much. This resulted in a smoky exhaust and a rapid build-up of carbon on the pistons. The uncertainties of this system were compounded by the variable quality of some of the early oils, which could affect how easily they flowed around the engine.

The answer was a pressurised system that pumped oil around the engine in a more controlled way. Such a system was successfully developed for a high-speed steam engine in 1890, but the manufacturers of internal combustion engines were slow to understand its application to them. From about 1903 some of the more sophisticated engines had a hybrid system whereby oil was pumped under pressure from the sump to key parts such as the main and big-end bearings, while the rest of the engine relied on splash lubrication.

One person who did understand about proper lubrication was Frederick Lanchester, and his 1905 20hp model was the first to be fitted with fully forced lubrication, operating at about 40lb/sq-in. Many of the other cars of the period operated at a pressure of just 5 to 10lb/sq-in, which offered far less protection to the engine.

THE VITAL SPARK

Having got the combustible mixture into the cylinder, it then had to be ignited. As early as 1807 the De Rivaz motor carriage used a system of electrical sparks to do this, and the principle was later applied in Lenoir's 1862 motor carriage and Benz's 1886 car. But the size, weight and reliability of these early systems were all serious issues, and they were also limited by their reliance on batteries or accumulators (the latter needing removing from the car and recharging every twenty-five hours or so – a heavy and messy business). There were also other problems. If you parked the car with the engine on the firing point, and forgot to switch the current off straight away, the ignition system would keep working and you would return to find your battery flat. More fundamentally, the faster the engine ran, the less current flowed through the system and the weaker the spark became.

As we saw, the early Otto engines, produced during the association of Daimler and Maybach with Otto's company, did not use electrical means at all. Instead they had an arrangement whereby a slide valve momentarily opened to expose the fuel mixture to a flame. Such an arrangement was never going to produce a high-revving engine. When Daimler and Maybach left them to set up on their own, their revised engine design used instead a patented 'hot tube' system to ignite the fuel, enabling their engine to rev up to an unprecedented 900rpm. This used a hollow platinum tube (experimentation showed that platinum was the only metal that did the job efficiently), one end of which went into the cylinder. The other end went into a variety of blowlamp, where it was heated by burning petrol or paraffin. The hot tube ignited the petrol-air mixture when it entered the cylinder. The idea of the hot tube was not, however, wholly original. One Leo Funk patented something similar in 1879, from which Daimler is thought to have drawn the inspiration for his 1886 model. Hot tube ignition was the dominant form of ignition on the earlier nineteenth-century internal combustion engines.

The hot tube system also had considerable shortcomings. First, the system gave little control over the timing of ignition – all that could be done in this respect was to move the burner closer to, or further from, the cylinder end of the tube to advance or retard the ignition slightly. Second, it had its own reliability problems, in that it was prone to being blown out by the wind – a growing problem as cars got faster. Third, it was a significant hazard having an open burner in the engine compartment, in close proximity to carburettors that were not immune to leaking petrol or petrol vapour. Putting out fires could become a regular part of the motoring experience. It is said that an early Daimler salesman, seeing his vehicle being consumed by flames, tried to persuade the prospective purchaser that this was simply a way of warming the vehicle up. Fourth, and perhaps the least of its problems, was that the burner or burners had to be lit to get the engine going, making a fast getaway from cold an impossibility.

The shortcomings of both systems led in 1895 to the English inventor and motoring pioneer F.R. Simms experimenting with a self-contained generator for a car. He had the Stuttgart electrical firm of Robert Bosch build a prototype of his low-tension magneto. As with many motoring developments, others, in this case Markus and Otto, had worked on the idea before. With the magneto system it was possible to retard the ignition to help starting, and advance it as the engine ran more quickly. One other major advance with it was that, by reducing the weight of its moving parts, it was able to accommodate a significant increase in the operational speed of the engine. The 1897 Lanchester was possibly the first car to be fitted with low-tension magneto ignition and, from about 1900, the hot tube system was rendered increasingly obsolete. F.R. Simms tried to get Daimler to adopt the Bosch system, but this fell through, for business and personality, rather than technical, reasons. Maybach came up with the compromise solution of fitting their cars with both hot tubes and electric ignition. Edward VII bought two dual ignition Daimlers, a model which eventually grew to a massive 10.4 litres and apparently required superhuman strength to turn over on the starting handle.

The low-tension magneto was in turn overtaken from about 1906 by the high-tension version. However, the magneto was expensive and difficult to manufacture, and provided a weak spark at low speeds, making starting difficult. Some luxury cars of this period were therefore fitted with dual systems, with a coil for starting and a magneto for normal running. Some even went as far as to have two completely independent ignition systems, each with its own set of sparking plugs.

Early sparking plugs had their own problems, in that the metal and porcelain parts of them tended to expand at different rates when heated, leading to leakage of pressure from the cylinder. As a minor historical aside, a racing cyclist called Albert Champion moved to the United States to become a chauffeur, where he went on to found the sparking plug company that so famously bears his name.

STAY COOL

The need to cool internal combustion engines was not universally understood by the early motor manufacturers, despite the very name 'internal combustion' providing something of a clue. The worst designs totally ignored the problem, whilst others (including the 1886 Daimler) made inadequate provision for versions of air cooling. Even where water-cooling was used, most did not at first use a sealed system with a radiator. Instead, they simply carried a large

More trouble keeping cool. Duray in his Darracq comes steaming to a halt in the 1903 Gordon Bennett race.

tank of water which was allowed to boil away as it performed its cooling function. This not only represented an unnecessary weight penalty for the vehicle; it also meant that it was reliant upon regular refills of water (though in the days of roadside horse troughs rather than petrol filling stations, this may not have been as great a problem as it sounds). By 1898 the Benz 'Comfortable' had at least gone so far as to introduce a water condenser to reduce water loss by this means.

The next step was to have a simple tube radiator to carry the cooling water. Some early models used the tubular frame of the car itself to carry the water, and there were other ingenious arrangements. The Clarkson and Capel radiator consisted of a helix of tinned copper wire spirally bound to the tubes carrying the water and then soldered onto it to increase the cooling surface. The 1908 Napier record-breaker at Brooklands had the radiator tubes forming the sides of the engine compartment. But none of these arrangements had an adequate surface to volume ratio to cool increasingly powerful engines. For many manufacturers up to and after the First World War the answer lay in the honeycomb radiator, in which the two ends of rounded tubes were expanded out into hexagons. These hexagons were then nested together into an arrangement that looked like a honeycomb, made watertight by being dipped in solder. The water could then run through the central part of the assembly and be cooled by the air stream blowing through the honeycomb. This gave a much greater surface area of metal exposed to the air in relation to the volume of water, and allowed a smaller quantity of water to dissipate much more heat. Pressurising the system had the effect of increasing the boiling point of the cooling water, but such systems were not in general use until after 1914.

On most early cars the water circulated round the cooling system by the natural process of convection, but, again, increased performance demanded something more effective, so mechanical means were introduced to pump the water around.

The other problem with water was that it could freeze in cold weather, cracking the engine block. This initially meant that the car either had to be kept in a heated garage or had to be drained overnight in cold weather. Knight advised Benz owners in 1902:

> It is absolutely necessary in winter to empty the tank by the tap provided for that purpose. The tap is very awkwardly situated, being almost in the middle of the car. A stick with a deep notch cut out at one end to fit the handle of the tap is very useful for the purpose.
>
> (Page 29)

Various additives were tried for their anti-freeze properties and found unsatisfactory, until a combination of ethylene glycol with other ingredients to counteract its corrosive qualities was found.

STOP!

With a horse-drawn vehicle the driver could (hopefully) rely upon the cooperation of the horses to bring the vehicle to a halt. No such possibility existed for the early motorist. The very first car brakes used the carriage and bicycle technology of the day. Hand-operated brake blocks would be applied direct to the solid tyres. These were not very effective, especially in the wet, and could damage the tyres. They were particularly lethal on hills because the front edge of the shoe on some models could wear down to be as sharp as a knife. If the car started to roll backwards downhill and the brake were applied, this leading edge could cut into the tyre and destroy it. Benz was one of the earliest to try a brake that operated on the transmission. The 1891 Panhard-Levassor combined a pedal-operated brake on the transmission and a hand lever that worked band brakes on the hubs, and this arrangement was also used by a number of other makers.

The brakes on the earliest cars tended to be contracting band brakes applied to the outside of either the brake drums on the rear axles or the car's drive shafts (or a combination of the two).

They were lined with what was supposed to be frictional material, to aid grip. In some cases this material was not as frictional as it might be (leather or wood) and, in the worst examples, the brakes were virtually useless. In others, the brakes were rather too frictional (in that they had a tendency to lock solid). Where the leather material in a clutch or brake lining lost its grip (for example, when it got wet) one practice was to put resin on it. Whilst this temporarily improved the grip, the solidified resin had later to be scraped off – a laborious process. Band brakes were gradually replaced by internal drum brakes, in which pivoted shoes inside a brake drum were forced open by a cam to press outwards against the drum.

Front wheel braking was rather more complicated in the days before hydraulically operated brakes, given the need for the front wheels to steer. A method of front wheel braking with cable-operated brakes was patented by T.G. Allen in 1906. This provided a system of pulleys that enabled both of the front wheel cable brakes to be operated, no matter in what direction the front wheels were pointing. However, the system turned out to have a fatal flaw in that, under some circumstances, it could lock up the steering. A few makers introduced forms of four-wheel braking before 1914, but even many of the 1914 Grand Prix cars still had rear-wheel braking only.

However, under the regulations agreed by the Local Government Board after the 1896 Act, road-going cars had to have two independent braking systems, either of which could lock both the wheels on one axle. This effectively meant the rear axle, which meant either fitting complete duplicate braking systems on the axle (which was expensive), or separately operated braking systems sharing a common brake drum. With the latter, if the first braking system failed from overheating the second system was also operating in an overheated brake drum and was itself likely to fail.

Transmission brakes were another option, but they tended to put excessive stress on the transmission and could be rather brutal in their braking effects.

One of the braking systems would be operated by a foot pedal and the other by a hand lever, but manufacturers differed as to which should be the brake for normal driving and which the emergency back-up. Those models using the hand brake as the main system presented something of a challenge for the driver, who also had to steer and operate the hand throttle at the same time.

The power of the brakes was severely limited by the tyre technology of the day, it being all too easy to lock up the wheels and lose control. Another limiting factor was the effectiveness of the friction materials used in the brakes. As we saw, wood and leather very soon became inadequate for the pressures being put on them. In 1901 Herbert Frood patented a braking material using layers of textile material impregnated with rubber. By 1908 textiles had been replaced by asbestos, to prevent charring, and his Ferodo Company was the market leader in its field.

Braking was a problem for the earliest cars, not least when the engine proved unequal to climbing a steep hill. I described earlier some of the primitive devices that were dragged along the ground to slow stagecoaches going downhill, and some of the early cars were fitted with equally primitive attachments for hills. The difference was that these were designed to stop the car running backwards, should it grind to a halt going up a hill, since the early brakes tended only to work when the car was going forwards. An 1898 edition of *The Autocar* carried an account of a trip from John O' Groats to Land's End (the one with which this book started) in a Daimler fitted with a *sprag* or *devil:*

> I also had at Mr Ellis' suggestion, a devil fitted beneath the car, the said devil being a long heavy rod, like a crowbar, which, whilst held up by a cord when not in use, could be let down on a steep incline, so that, should the car stop, it would dig into the ground and prevent the vehicle running back.
>
> (Hough, page 111)

The hope was that the sprag would work immediately, digging into the ground and holding the car still. If it did not, the danger was that the car would start to roll backwards until it was stopped violently, possibly pulling the axle off the springs or even causing the car to use the sprag as a vaulting pole and ending upside down.

LET THERE BE LIGHT

The earliest cars rarely went out at night. One reason for this may have been that they were dependent upon the same oil lamps and candles that were used to light the way for horse-powered transport. These were not even adequate for horse-drawn speeds and, as cars got faster, a much better system was needed. The Lucas Company, which was already a leading manufacturer of bicycle lights, went into the development of lamps for car use. Oddly enough, prior to 1907 there was not even a law requiring cars travelling during the hours of darkness to have proper lights, sufficient to show the driver the road ahead. All that the 1896 Act required was that, from one hour after sunset to one hour before sunrise, they should mark their presence to others by displaying a white light in the direction of their travel and a red one facing in the opposite direction. However, most manufacturers soon worked out that headlights were more than just an optional extra.

Since 1849 Britain's lighthouses had been lit by acetylene, a hydrocarbon gas produced by the action of water on calcium carbide, which burned with a brilliant white light. Before 1900 these were adapted for use on cars, and were steadily improved by the use of mirrors to intensify the beam, until it was said that they were bright enough to enable a car to travel at up to 40mph at night on a good road. There was even an anti-dazzle device developed for them in 1906. However, they were not ideal, requiring more attention than electric lighting, having the potential like all naked flames to be blown out and reach high temperatures (the flame in an oxy-acetylene torch reaches 2,800 degrees centigrade). The gas is also very explosive when under pressure and the process of producing it was messy and smelly (which led to the use of bottled acetylene in preference to an on-board generator). Even then, their performance did not keep pace with the development of cars. When S.F. Edge made his record-breaking twenty-four-hour run at Brooklands in 1907 he described night driving with acetylene lights as being 'like the skipper of a ship at sea without a compass'. He had to have the course lined with red lanterns at ten yard intervals to help him find his way.

A gleaming display of acetylene lighting on this Peugeot.

Electric lighting became a possibility with the development of the engine-driven dynamo soon after 1900, but they were not particularly efficient until about 1910. The 1911 Cadillac was the first model to have electric lighting as standard. By the First World War electric lights still tended only to be found at the luxury end of the car market, and acetylene lighting was still the norm.

WINDSCREENS

…dangerous abominations, unnecessary and in the way during fine weather and completely useless in the rain.

(An early view of windscreens, quoted in Roberts)

As we saw, one of the deterrents to improving weather protection on the earliest cars was the fact that a stone hitting the windscreen could fire deadly sharp shards of glass in the direction of the driver and passengers. A solution to the problem seemed to be in sight in 1905 when J.C. Wood patented a system for bonding two thin sheets of glass to a thin central skin of celluloid, so that the windscreen crazed rather than shattered upon impact. Unfortunately, having patented it, Wood found he could not manufacture it, and the patent lapsed. It was only revived after 1910, once improved glues had been developed and the Triplex Company had been founded. Even then, they had the problem that the early celluloids tended to go opaque over time.

Not all makers opted for safety glass immediately. It took until 1927, when Henry Ford was cut by flying glass in a car accident, before all Fords were fitted with safety glass. None of these windscreens would have had the benefit of windscreen wipers as we know them, at least within the time period in which we are interested. The 1904 Gardner Serpollet steam car had one which was hand-operated, but the first patent for a mechanical windscreen wiper, driven by the partial vacuum in the induction pipe of the engine, was taken out by W.M. Folberth in 1920. The electric version came even later, in a 1923 American patent. Its main benefit was seen to be to reduce the need for the chauffeur to open his window.

GEARS

On the earliest cars belt drive was common and changing gear involved either moving the belt from one pulley to another, or disengaging one belt and selecting another. The Benz Comfortable of 1898 had three separate levers for selecting the right gear ratio. Newcomb and Spurr summarise the advantages of belt drive as being cheapness, simplicity, quietness and the ability to survive sudden changes of speed without risk of damage. Reverse gear could even be used as a brake with belt drive. Their disadvantages were that they tended to slip; they stretched, with the result that constant adjustment was needed; and they took up a lot of space, since the longer the belt, the better its performance. As engines grew more powerful and higher-revving, the days of belt drive were numbered.

Friction drive involved a metal disc on the end of the driveshaft from the engine, which transferred its power to another moveable wheel placed at right-angles. For a low gear the wheel was located near the centre of the metal disc. For a higher gear it was moved out towards the faster-moving outer area of the disc. This gave an infinitely variable speed or torque for a constant engine speed, and reverse was available by moving the wheel to the opposite side of the disc. As long as there was no slippage between disc and wheel rim the system was relatively smooth and silent, but it needed a light touch from the driver. It was all-too-easy to wear flat spots on the wheel, which led to what was described as 'a horrible grating, whining noise from the drive'. It, too, was limited in the amount of power it could handle, though it lingered on in some forms of light car until after the First World War.

Minerva was a Belgian manufacturer. Patronised by many royal families, they were in production between 1899 and the outbreak of the Second World War. This 5,746cc special dates from 1918.

To overcome these shortcomings various forms of mechanical gearboxes were tried. Some of the early mechanisms operated in the open air, with only the occasional blob of grease for lubrication, and their operation was sometimes far from smooth. Design problems were compounded by the poor quality of some early steels and bad engineering. Gear teeth wore, bent or broke, leading to noisy gear-changing. All in all, gear-changing on the earliest cars was not an easy or pleasant business, and one of the reasons road cars went for larger and larger engines was for the flexibility it gave (and hence the reduced need to change gear). The publicity stunts staged by Rolls-Royce and Napier, of driving from London to Edinburgh in top gear only, had a serious point to make.

TRANSMISSIONS

Many cars had chains as part of their transmission until well into the twentieth century. However, they took a lot of maintenance, wore quickly and grew steadily noisier. According to Knight, 'a pair of Brampton chains, if well looked after, should run 2,500 to 3,000 miles.' It is claimed in some quarters that the first shaft-driven car was the Autocar, built in the United States in 1901, though many in Europe would say that Renault's first cars pre-date that. Moreover, Renault's attempted 'patent' of shaft drive was found not to have been his invention, so the technology must have pre-dated even him. By 1910 chain drive was obsolete on all but larger vehicles, and a driveshaft operating through a differential gear became the norm.

The differential compensated for the fact that the driven wheels went at different speeds as the vehicle rounded a corner. Without it (and some cheap cars actually went without it) tyres were scrubbed and wore out even more quickly than usual. Differential gears have ancient origins – the ancient Greeks and Chinese are thought to have understood the principle of them and, as we saw, Leonardo da Vinci drew a version of one. In modern times the idea was patented by one O. Pecquer in 1827, and they were used on most road-going powered-vehicles from about 1840 onwards. The main drawback of the differential is that, if one driven wheel hits a slippery patch, it can take all the drive and spin quickly, whilst the other one scarcely moves at all. Various attempts were made to produce a 'limited slip' differential that would prevent this happening.

One problem for early cars was preventing the rear axle from twisting (for example, from the effects of torque, or if one wheel hit a bump, causing the rear wheels to steer the car). The simplest way to attach the back axle was to fix it to the leaf springs alone (as the 1894 Renault was the first to do). But this on its own only worked for lighter, less powerful vehicles, so torque rods were fitted on some models to stabilise the rear axle. Others used a torque tube – a rigid tube in which the propeller shaft was carried, linked rigidly to the differential housing at one end and by a hinge to a cross member of the frame at the other.

The added complications of putting the drive through the steered front wheels did not deter early inventors. No less than 100 patents for it were registered by 1908, though how many of them would have worked is another matter. None of them found their way into a production car before 1914, though some pre-war American racing cars had front-wheel drive.

CHASSIS DESIGN

Given that so many of the early car manufacturers came from the cycle industry, it is no surprise that, up to the turn of the century, the chassis frames of smaller cars were built on bicycle principles, with round tubes brazed together. Larger cars followed coachbuilding technology, with reinforced wooden frames. Pressed channel steel largely replaced this from about 1903. Engines and gearboxes were generally on a separate sub-frame. As cars got longer and heavier the issue of body flexing became more prominent. Some designers did not worry too much about stiffening the chassis, using it in effect as part of the suspension, but the introduction of side doors made stiffening the chassis more of an imperative, since the sagging of the body would cause the doors to jam.

SUSPENSION

The experience of carriage and locomotive manufacturers meant that, in this area at least, the early car builders had a good deal of information to draw on. Steel springs for coaches were being experimented with as early as the mid-seventeenth century, and were in general use by about 1790. They helped achieve marked reductions in journey times. Elliptical springs were invented by Obadiah Elliott in 1804, which brought about a further increase in average coach speeds. The very earliest cars, if they had suspension at all, tended to use half- or fully elliptical springs. But the fully elliptical springs in particular took up too much space and made the cars sway too much, so a wide variety of different options, based on leaf springs, were tried in their place. Some – the Kanstatt-Daimlers, the American Brush and the 1903 Vauxhall, for examples – used coil springs, but the technical challenges of installing them and the difficulties of manufacturing them prevented their widespread use.

The need for shock absorbers never quite affected coach manufacturers in the way it did car makers. So there was little coachbuilding experience to draw upon. In the 1904 Gordon Bennett race Thery won in his Richard-Brasier, beating Jenatzy's Mercedes by nearly eleven minutes. Part of the difference was put down to the fact that Thery's car was fitted with shock absorbers, whereas Jenatzy's was not. They are first recorded in the context of the Mors' victory in the Paris–Madrid race of 1903, but relatively few pre-1914 cars were fitted with them.

STEERING

The earliest car manufacturers were apparently ignorant of earlier developments in steering. George Leckensburger of Munich had in 1816 devised a system which dealt with the fact that when a vehicle turned, the inside and outside front wheels needed to turn through circles of different sizes. Failure to do so could cause scrubbing of the tyres and other problems. Leckensburger solved the problem by having two separate pivoted stub axles at the ends of a fixed front axle. These were linked by a tie rod that ensured that each wheel had the correct turn-in on a bend. This system was patented in 1818 by Rudolph Ackermann, whose name it inherited, but was lost for nearly eighty years. Benz, Bollee and other pioneers were unaware of its existence. Benz, worried that a crude front axle that pivoted from the centre would be subject to violent 'kicking' if one wheel hit a pothole, got round the problem by having just a single steered front wheel on his first car. Others had to re-invent the Ackermann system.

Tiller steering was the norm on many of the earliest cars, but was dangerous at higher speeds. Road shocks were easily transmitted through it and could throw the tiller out of the driver's hands and the car out of control. It was partly responsible for one of the first fatalities in motor racing when, during the Course de Perigueux in 1898, the Marquis de Montaignac turned to wave to a car he was passing. The tiller kicked as he did so, causing him to swerve, forcing the other car off the road. Hearing the crash, the Marquis made the mistake of looking back to see what had happened, causing him to lose control again. He in turn rolled his own car, killing himself and his mechanic. Steering wheels began to replace tillers from about 1897, though Lanchester redesigned and improved the tiller mechanism, retaining it long after other makers.

A Vauxhall Viper, dating from 1913, with a massive 11,778cc engine.

Steering columns tended to be vertical on the earliest cars but, following the example of the 1901 Mercedes, became increasingly raked as cars became lower and more streamlined. The gearing on the steering of this period was generally low, making it positive but very heavy to operate. One revolution of the steering wheel was normally enough to go from lock to lock. One other feature of some early steering systems was the use of chains, the failure of which could have catastrophic results. Knight is unambiguous in advising would-be purchasers of cars in 1902 that 'a car in which a chain forms any part of the steering gear should be avoided.' But the prize for eccentric steering systems must go to the 1904 Turicum, a Swiss car that was steered by two foot pedals.

TYRES

The result of the race only served to strengthen the unfavourable opinion I had formed of the pneumatic tyre, which might possibly be filled with compressed hay or straw, but would never be serviceable when filled with air.

(Emile Levassor, in a letter of 2 August 1895)

In practical tests of both wire wheels and pneumatic tires and wood wheels and solid rubber tires, it has been demonstrated to the satisfaction of the designer that the latter are far more desirable and endurable in many ways than the former, and present a more satisfactory appearance, and all annoyances due to punctures are done away with.

(*Scientific American*, 13 May 1899)

The earliest cars shared the bone-shaking solid-tyred ride of many of their horse-drawn counterparts. In addition to being very uncomfortable at anything above 20mph, the tyres were noisy, expensive and inclined to part company with the wheel. Most manufacturers gave up solid tyres in the early 1900s, although the 1920s Trojan still used them, offering pneumatic tyres as an optional extra. The first important breakthrough was made as long ago as 1839 when a process known as vulcanisation was discovered that produced a rubber durable enough for use in tyres. It was invented independently, in the same year and on both sides of the Atlantic, by Thomas Hancock in Britain (brother of the steam carriage pioneer Walter) and by Charles Goodyear in the United States.

Although it was 1895 before the first pneumatic tyres were fitted to a car, the technology was known and potentially available for half a century before that. A Scot named Robert Thomson had taken out a patent in December 1845 for an 'improvement in carriage wheels which is also applicable to other rolling bodies' – the original pneumatic tyre. *Mechanic's Magazine* in 1847 reported seeing 'a brougham with silent wheels, so silent as to suggest a practical inconsistency of the most startling kind between the name and quality of the thing.' These new devices could be 'inflated with air to any degree of tightness required'. That same year, the patent was sold on to a company called Whitehurst, but the tyres proved to be too expensive and difficult to fit (they required seventy bolts to hold each one on the wheel) and the idea fell into obscurity.

Forty years later and the idea was revived by another Scot. John Boyd Dunlop was at that time practicing as a vet in Dublin. His son's tricycle was making ruts in the lawn and, quite oblivious to the earlier patent, Dunlop hit upon the idea of wrapping lengths of garden hose around the wheels as tyres. Initially he filled them with water, but the family doctor suggested he use air instead. Dunlop patented the idea in 1888 but, two years later, *Sport and Play* magazine revealed the existence of Thomson's earlier patent. This invalidated Dunlop's claims and opened the market to rival manufacturers, but this did not prevent Dunlop's company, the Pneumatic Tyre and Booth's Cycle Agency Ltd, which only later took the name of its founder, enjoying great commercial success. As its name implies, the primary market for the company's pneumatic tyres was the bicycle. They were still difficult to use – one contemporary likened repairing them to 'an operation for appendicitis with complications' – but their overwhelming success in bicycle racing made their commercial progress irresistible.

Dunlop himself saw few of the financial benefits of his invention. A biographer said of him that 'his career was a classic example of an inventor largely devoid of entrepreneurial skills… his role in the foundation of the large British rubber manufacturing company which now bears his name was a limited one.' Dunlop died in 1921 leaving an estate of less than £10,000 gross.

But it was not Dunlop who pioneered the use of pneumatic tyres on the motor car. The brothers Andre and Edouard Michelin owned a factory near Paris making a variety of rubber goods. One day an English cyclist arrived at their premises asking for his punctured Dunlop tyre to be repaired. The brothers were struck by the difficulty of doing so and set out to design something better. In 1891 they patented a tyre which fastened to the wheel by means of a steel band and bolts, and with a separate inner tube. This enjoyed considerable success in cycle racing, largely because it was far quicker than the competition to change when punctured. From there the company diversified into pneumatic tyres for horse-drawn vehicles and motor cars. One of their first experiments was with one of the Peugeots entered in the 1895 Paris–Bordeaux–Paris race. The car carried twenty-two spare inner tubes and used them all, suffering a total of fifty punctures, but it did at least finish the race. About half the entrants in the next year's Paris–Marseilles race used Michelin pneumatic tyres, and the brothers collected every tyre used in the race and analysed them for weaknesses. One lesson they soon learned was that bicycle tyre technology was not applicable to the stresses experienced by a car tyre. It was replaced by the Doughty process, in which the entire tyre was vulcanised, between moulds and under pressure.

Until 1903 cars were required by law not to have tread patterns (with the aim of minimising road damage). Tread patterns would in any event have been difficult to produce with the manufacturing technology of the day and would soon have been worn away by those cars with brakes acting directly onto the rear tyres. Once tread patterns were legal, experiments were also made with various forms of metal studs, chains and other techniques to reduce skidding, but these generally shortened the life of the tyre and did an unacceptable amount of damage to the roads.

Punctures were the nightmare of early motorists. Apart from the delay and inconvenience, the cost of each repair in America often came to between $5 and $10. They became a little less nightmarish with the introduction of the Stepney rim, which fitted next to the punctured tyre. Detachable rims would not be long in following, and may have been accelerated in their development by a ruling for the 1906 French Grand Prix that all repairs to the racing cars (including punctures) had to be carried out by the driver and his mechanic. Changing beaded-edge tyres in the days before removable wheels was a very hard business indeed. In a contest in London in 1905 an employee of Wolseley and a specialist in his trade, working under competition conditions, took five minutes and four seconds to change a tyre. Some of the leading teams in the Grand Prix – including Brasier, FIAT, Renault and Itala – therefore turned up at the race with detachable rims. Cars with these came first and second in the race.

But, for all the development work done by Michelin and others, tyres remained an expensive and problematic part of motoring in this period. C.S. Rolls estimated that, in 1900, tyres on his (admittedly powerful) car cost him 2d a mile. Even on a light car, driven carefully, you would be lucky at that time to get 2,000 miles from a set of tyres. 1,000 miles would be more typical, and it would be unusual to complete a long journey without at least one puncture.

BODYWORK

We looked earlier at the coachbuilding tradition and some of the anomalies that it produced in early car design. Many of the bodies on these look to have been positively designed to impede the progress of the car through the air. In the years immediately preceding the First World War, greater interest began to be taken in the idea of streamlining, both for aesthetic purposes and to improve the efficiency of the car. Aerodynamics had been studied since the mid-nineteenth century in relation to ships and the emerging field of aviation, though wind tunnels for more scientific evaluation were not available until after the First World War. Frederick Lanchester, in particular, was a car designer with considerable expertise in aerodynamics. However, as Georgano points out, much of the aerodynamic theory applied to the car body design of that period was based on a misconception. They believed that an elongated teardrop shape was the most aerodynamically efficient, following the example of fish. This principle was applied not only to cars but to submarines and Zeppelins. It is wrong on two counts; first, droplets of liquid do not actually form a teardrop shape in freefall. Second, the teardrop shape is of no aerodynamic benefit to a car moving close to the ground. The result in some cases was an elaborate attempt to 'streamline' the rear of the car (to no real practical effect) combined with some very un-aerodynamic frontal treatment (such as large vertical windscreens and radiators).

AND SOME OTHER PARTS

In this section we look at the origins of some of the other features we take for granted on a modern car. Much of the heating provided on early cars may have come from the unintentional effects of their inadequate cooling systems. Wolseley installed hot water heating beneath the rear seats of their 1902 model, but the idea did not surface again until Lanchester experimented with heat from the exhaust. In the meantime, many cars carried 'carpet carriage warmers', giant metal hot water bottles.

Then there was: 'A neat idea…the arrangement of a mirror above the front screen, which, in conjunction with the light in the back panel, shows any traffic approaching from the rear', which appeared in 1911. We call it the rear-view mirror.

An 8.2-litre Lorraine-Dietrich from 1905, the year in which they began car manufacture. The marque would later win the Le Mans twenty-four-hour race two years running.

As for in-car entertainment, the first car radio (or 'wireless telegraphy installation') made its appearance in 1910. But two factors prevented it from becoming an instant success; first, the apparatus took up most of the rear seats and, second, there were no British radio stations for it to receive until 2LO began broadcasting until 1922.

The first cable-operated direction indicators appeared in 1913. There were two varieties; one was a drum with an arrow that revolved in it, and the other a white celluloid hand that stuck out to point to where you were turning.

Car theft was an issue early on. Henry Ford used to chain his first vehicle to lamp posts to prevent people taking it for joyrides. As cars became more common, and more desirable objects to own, so the possibility of their theft increased, so manufacturers began fitting them with locks. Whilst they prevented thieves from climbing in, they also prevented legitimate passengers from falling out. Mulliner began fitting double catches on the doors so that 'if imperfectly closed, say by a theatre attendant, they will not fly open.'

Finally, some motor cars with features not found today on your average car: the Maxim gun and the electrified body. From the very start military applications were part of the development of the car. Isaac de Rivaz's 1813 attempt at an internal combustion-engined vehicle was inspired by ideas of moving military materials, as were Nicholas Cugnot's earlier experiments with steam. Steam traction engines were in use at the Woolwich Arsenal in the middle of the nineteenth century, and forty-six of these vehicles saw military service in the Boer War. Some of the country's largest armament companies, such as Vickers (who bought the Wolseley Company) and Armstrong Whitworth, got involved in the motor industry.

At the Automobile Club's show in June 1899 Frederick Simms exhibited a range of motorised military vehicles. The light one, or 'motor scout', was a quadricycle equipped with a Maxim gun. The 'war motor car', which anticipated the tank by almost two decades, was a heavier, fully armoured vehicle with rams front and rear and two Maxim guns carried on revolving turrets. It was steered by mirrors to prevent the driver exposing himself to enemy fire. An additional feature was that the entire body could be electrified to shock anyone touching the outside (one wonders how well those inside were protected from the shock?).

In 1904 the War Office offered a prize of £1,000 for a traction engine suitable for military purposes. The specification effectively ruled out conventional steam traction engines since the machine had to be capable of travelling forty miles at an average of 3mph and a maximum of 8mph without taking on fuel. The successful vehicle, built by Hornsby-Ackroyd, looked like a conventional traction engine but was powered by an oil engine. A Motor Volunteer Corps (later the Army Motor Reserve) was set up in Britain before the First World War. Come the hostilities, the cars were conscripted for conversion to ambulances or armoured cars.

The American Army used armoured Cadillacs in their hunt for insurgents on the Mexican border. But perhaps the most famous example of the military use of ordinary motor cars was to come in the early stages of the First World War when the French army mobilised 600 Parisian taxis to carry five battalions of infantry to the nearby battlefront to ward off the imminent threat to Paris itself. For this excursion into the war zone, the cabbies were allowed to add 27 per cent to their normal meter fare.

But even this had a British precedent. On St Patrick's Day 1909 the Automobile Association staged a publicity stunt involving the transporting of an entire battalion of Guardsmen from London to Hastings as a demonstration of the potential for motor vehicles to move large volumes of military materiel about the country. They even arranged to transport the soldiers' bulky equipment by persuading one of London's largest taxi firms to strip the bodies off thirty of their cabs and convert them into flat-bed lorries for the day. The event may have opened the eyes of the British Army to the motor age, but across the Channel the event was studied in equal detail and with equal interest by the German High Command.

One more immediate consequence of this event was the fitting of chinstraps to service caps to prevent them being blown away in the slipstream. The guards had previously had to resort to the unsoldierly practice of tying them on with their handkerchiefs.

EPILOGUE

PRESERVATION – AND A LOST OPPORTUNITY

By 1912 most of the first generation of motor cars had reached the end of their working lives, and were technologically outmoded. Far-sighted enthusiasts, including Edmund Dangerfield, the managing director of Temple Press, the proprietors of *The Motor* magazine, saw their historical importance and feared that they would be lost to future generations if action were not taken. A Board of Trustees was set up under the Duke of Teck with the aim of preserving a selection of these vehicles. Leading museums were approached, but showed very little interest – the Science Museum had difficulty in accommodating just three of them – so the Trustees were forced to take independent action.

A collection of thirty-nine cars, most of them dating from before 1900, was assembled and an exhibition opened in June 1912. This was moved in March 1914 to the Crystal Palace at Sydenham where it remained until the building was requisitioned for the war effort and the collection had to be broken up. A number of them were simply left in the open, on a piece of waste ground near Charing Cross Station, where they deteriorated and gradually disappeared. Thus was lost an important part of Britain's motoring heritage.

BIBLIOGRAPHY

Aldridge, John and Morris, Stephen, *British Buses Since 1900* (Ian Allan, 2000)

Arnold, James, *All Drawn by Horses* (David & Charles, 1979)

Autocar

Bak, Richard, *Henry and Edsel: The Creation of the Ford Empire* (Wiley, 2003)

Barker, Theo and Gerhold, Dorian, *The Rise and Rise of Road Transport 1700-1990* (The Economic History Society, 1993)

Boddy, William, *The History of the Brooklands Motor Course* (Grenville, 1957)

Boddy, William, *Brooklands Giants* (Foulis, 1995)

Box, Rob de la Rive, *The Complete Encyclopedia of Vintage Cars (1886-1940)* (Rebo Netherlands, 1998)

www.brighton-early.com

www.brooklands.org.uk

Canning, John, *100 Great Modern Lives* (Souvenir Press)

Cauntier, C.F., *Science Museum: Handbook of the Collection Illustrating Motor Cars* (Part 2) (HMSO, 1959)

Cauntier, C.F., *The Light Car* (HMSO, 1970)

Church, Roy, *The Rise and Decline of the British Motor Industry* (Macmillan, 1994)

Cooke, Stenson, *This Motoring (being the romantic story of the Automobile Association)* (Automobile Association, undated)

Georgano, Nick, *The Beaulieu Encyclopedia of the Automobile* (TSO, 2000)

Georgano, Nick (editor), *The Beaulieu Encyclopedia of the Automobile - Coachbuilding* (TSO, 2001)

Glancey, Jonathan, *The Car: A History of the Automobile* (Carlton, 2003)

Harrison, Ian, *The Book of Inventions* (Cassell, 2004)

Hattersley, Roy, *The Edwardians* (Little, Brown, 2004)

Holloway, Hilton and Buckley, Martin, *The A-Z of Landmark Cars* (Carlton, 1998)

Homans, James E., *Self-propelled Vehicles* (Audel & Company, 1908)

Hough, Richard (editor), *Motor Car Lover's Companion* (George Allen & Unwin, 1965)

Jellinek-Mercedes, Guy, *My Father Mister Mercedes* (G.T. Foulis, 1961)

Johnson, Howard, *Wings over Brooklands* (Whittet, 1981)

Journal of the Illinois State Historical Society (spring 2003), *The Race of the Century: 1895 Chicago*

Karslake, Kent and Pomeroy, Laurence, *From Veteran to Vintage* (Temple Press, 1956)

King, Peter, *The Motor Men* (Quiller, 1989)

Knight, John Henry, *Light Motor Cars and Voiturettes* (1902)

Lee, Charles E., *The Early Motor Bus* (London Transport, 1974)

Newcomb, T.P. and Spurr, R.T., *A Technical History of the Motor Car* (Adam Hilger, 1989)

Nicholson, T.R., *Passenger Cars 1905-1912* (Blandford, 1971)

Nicholson, T.R., *The Birth of the British Motor Car* (Volume 1: A New Machine 1769-1842) (Macmillan, 1982)

Nicholson, T.R., *The Birth of the British Motor Car* (Volume 3: The Last Battle 1894-97) (Macmillan, 1982)

Nixon, St John C., *The Antique Automobile* (Cassell, 1956)

Outing magazine

Oxford Illustrated Encyclopedia of Invention and Technology (OUP, 1992)

Parry, David, *English Horse Drawn Vehicles* (Warne, 1979)

Pettifer, Julian and Turner, Nigel, *Automania* (Guild, 1984)

Roberts, Peter, *Veteran and Vintage Cars* (Hamlyn, 1963)

Robson, Graham and Ware, Michael, *Illustrated Guide to Classic British Cars* (Abbeydale, 2000)

Ruiz, Marco, *The History of the Automobile* (Arnoldo Mondadori S.p.A. Milan, 1984)

Schroeder, Joseph J. Jr. (Editor), *The Wonderful World of Automobiles 1895-1930* (Digest Books, 1971)

Scientific American

Scott-Moncrieff, David, *Veteran and Edwardian Motor Cars* (Batsford, 1955)

Setright, L.J.K., *Drive On! A Social History of the Motor Car* (Palawan, 2002)

Stevens-Stratten, *British Lorries 1900-92* (Ian Allan 1992)

Thorold, Peter, *The Motoring Age: The Automobile and Britain 1896-1939* (Profile, 2003)

The Times

Troubridge, Lady and Marsall, Archibald, *Lord Montagu of Beaulieu: A Memoir* (1985)

Turner, E.S., *An ABC of Nostalgia* (Michael Joseph, 1984)

Tyler, John, *Morgan, the Cars and the Factory* (Crowood, 1993)

Villard, Henry Serrano, *The Great Road Races 1894-1914* (Arthur Barker, 1972)

Wilkes, Peter, *An Illustrated History of Traction Engines* (Spurbooks, 1974)

INDEX